THE POWER
OF ONE

THE POWER OF ONE

How I Found the Strength
to Tell the Truth and
Why I Blew the Whistle
on Facebook

FRANCES HAUGEN

LITTLE, BROWN AND COMPANY

New York   Boston   London

I have done my best to recount this story as accurately and as truthfully as possible—it is true to what I believe happened. I have changed some names and some descriptions to protect the privacy of others, and I have reconstructed dialogue to the best of my recollection and reordered or combined the sequence of some events. Others who were present might recall things differently, but this is what I remember.

This book recalls my experience whistleblowing and mine alone. Any discussion of the logistics of whistleblowing is not intended to be construed as instructions for the process of whistleblowing. I recognize I was part of a moment in history, and I am attempting to help others understand this moment and the context around it. I am not a lawyer and this book is in no way a substitute for the advice of a competent attorney admitted to practice in your jurisdiction. If you attempt to report corporate wrongdoing, your ultimate success or failure will be the result of your own efforts, your particular situation, and innumerable other circumstances beyond my knowledge and control.

The opinions expressed in this publication are those of the author and do not necessarily reflect the official policies or positions of Facebook, Google, Yelp, or Pinterest.

Little, Brown and Company
Hachette Book Group
1290 Avenue of the Americas, New York, NY 10104
littlebrown.com

First Edition: June 2023

Little, Brown and Company is a division of Hachette Book Group, Inc. The Little, Brown name and logo are trademarks of Hachette Book Group, Inc.

The publisher is not responsible for websites (or their content) that are not owned by the publisher.

The Hachette Speakers Bureau provides a wide range of authors for speaking events. To find out more, go to hachettespeakersbureau.com or email hachettespeakers@hbgusa.com.

Little, Brown and Company books may be purchased in bulk for business, educational, or promotional use. For information, please contact your local bookseller or the Hachette Book Group Special Markets Department at special.markets@hbgusa.com.

ISBN 9780316475228
Library of Congress Control Number: 2022951277

Printing 1, 2023

LSC-C

Printed in the United States of America

This book is dedicated to you, the reader. With each page you turn, you expand the community of people who can help build the brighter future of social media that we all need and deserve. The path forward is about more than fear and despair. This is only the start of a conversation that will change the world.

Contents

CHAPTER 1

State of the Union

The exercise of power is determined by thousands of interactions between the world of the powerful and that of the powerless, all the more so because these worlds are never divided by a sharp line: everyone has a small part of himself [*or herself*] in both.

—Václav Havel, *Disturbing the Peace*

Don't worry," the boy said, looking up at me as we rode the elevator in the United States Capitol. "I've been doing this for a while and even I get butterflies sometimes." His words startled me out of my controlled breathing, a calming exercise I've found helps center me when I feel anxious. From the moment we had exited the White House and boarded the shuttle that whisked us to the Capitol, I felt as if I had stepped onto a steadily building escalator of anxiety and I didn't know how to get off. It was March 1, 2022, the evening of President Joe Biden's first State of the Union Address. Only five days before, Russia had invaded Ukraine. It occurred to me that the speech would draw even more attention than usual, as people wondered whether Biden might declare war on Russia. My heart was racing.

I glanced down at the boy, Joshua Davis. He wore a dapper dark blue suit, sapphire-blue tie, his blond hair parted on the side. The bespectacled thirteen-year-old emanated the poise of a seasoned ambassador. Which, in a way, he was. Diagnosed with diabetes as a baby, by the time Joshua was in kindergarten he had become something of a national spokesman on behalf of people with the disease. He had most recently been calling on the drug companies to make insulin affordable to all who needed it. Joshua was clearly comfortable at the center of attention, and he was clearly perceptive, as he could see that I most definitely was not at ease.

I had entered what became a spotlight just six months earlier, blowing the whistle on Facebook in a very public way, and testifying to Congress and elsewhere about the many routes by which the platform had become a source of misinformation and a spark plug for political violence. The company knew it was happening, but they prioritized profits over public safety.

The irony was not lost on me that I was now being reassured by a junior high student one third my age. I had a flash of a thought of how different we were: Joshua had spoken before the Virginia General Assembly at the age of four, urging them to pass a bill making schools safer for kids with Type 1 diabetes. When I was four years old, I was building wooden boxes only a mother could love, with real saws and hammers at my Montessori preschool. Up until six months before, when I revealed my identity on *60 Minutes*, I had spent my entire life avoiding the spotlight, to the point of having eloped to a Zanzibarian beach for my first marriage. In the fifteen-plus years since college, I've had maybe two birthday parties. My mind is wired to think in terms of data and spreadsheets, and according to my rough estimate, Joshua had been in the public eye for 70 percent of his life, whereas I had only been in the spotlight for less than 1.5 percent of mine.

We were among a handful of people that night who had been invited as guests of the First Lady. Being invited to sit in the First Lady's box meant the president of the United States would cite each of us in his address, humanizing symbols of his agenda. I had been invited because I was "the Facebook whistleblower." I had extracted 22,000 pages of documents from inside the social media company where I had worked on the Civic Misinformation team and then with Counter-Espionage. I not only worked to ensure that all of the technical and terrible facts in those documents made it into the public sphere, but by the time of the State of the Union, I had spent months on the road to make sure the public understood what they really meant.

I had made it through my public appearances thus far, including a debut on *60 Minutes* and testifying to a string of congressional and parliamentary committees around the world, by focusing on the presentation of the substance of the documents. I clung to an anxiety-relieving conceit that

I was, as a friend coached me, "just a conduit for the documents." My purpose was to provide clarity and context; my physical presence was incidental to that. It wasn't about me, it was about the information the world needed to know. This State of the Union, though, felt different. For this appearance, my purpose more or less was just *to be there*. To be *looked at*. When the President of the United States gave me my cue, I was to stand before the nation, before the world, and just be seen.

Shorn of my protective mantra, my heart was racing. "Thank you, you're so kind," I said to Joshua as we emerged into the marble-lined corridors of the Capitol and headed toward the balcony of the House of Representatives Chamber.

I had begun this journey the year before, when I submitted what I believed to be documents of immediate and immense public interest to the Securities and Exchange Commission as part of a whistleblower complaint. In my complaint, I detailed the myriad ways Facebook had, over and over again, misled—over and over again *failed to warn* the public about issues as diverse and as dire as national and international security threats, the Facebook algorithms that drive political party platforms, and the fact that Facebook had knowingly been harming the health and well-being of children as young as ten years old—all for the sake of profit. The complaint, rooted in the documentary evidence, made clear that Facebook was endangering the world and that the company was stuck in a downward feedback loop that would only get worse unless and until the public was made aware and it was compelled by regulatory intervention to change.

Facebook had been getting away with so much because it runs on closed software in isolated data centers beyond the reach of the public. Facebook realized early on that because its software was closed, the company had the upper hand to control and shape the narrative surrounding these and so many other problems it created. If there was no external awareness of the problems, if there was no awareness of the truth, then there would be no external pressures to deal with those problems. Software is different from physical products because the user can see its results only on a screen. We cannot see into the vast tangle of algorithms that produce that output—even if those algorithms exact a crushing, incalculable cost, such as unfairly influencing national elections, toppling governments, fomenting genocide, or causing a teenage girl's self-esteem to plummet, leading to another death by suicide.

One of the questions I was often asked after I went public was, "Why are there so few whistleblowers at other technology companies, like, say, Apple?" My answer: Apple lacks the incentive or the ability to lie to the public about the most meaningful dimensions of their business. For physical products like an Apple phone or laptop, anyone can examine the physical inputs (like metals or other natural resources) and ask where they came from and the conditions of their mining, or monitor the physical products and pollution generated to understand societal harms the company is externalizing. Scientists can place sensors outside an Apple factory and monitor the pollutants that may vent into the sky or flow into rivers and oceans. People can and do take apart Apple products within hours of their release and publish YouTube videos confirming the benchmarks Apple has promoted, or verify that the parts Apple claims are in there, are in fact there. Apple knows that if they lie to the public, they will be caught, and quickly.

Facebook, on the other hand, provided a social network that presented a different product to every user in the world. We—and by we, I mean parents, children, voters, legislators, businesses, consumers, terrorists, sex-traffickers, *everyone*—were limited by our own individual experiences in trying to assess *What is Facebook,* exactly? We had no way to tell how representative, how widespread or not, the user experience and harms each of us encountered was. As a result, it didn't matter if activists came forward and reported Facebook was enabling child exploitation, terrorist recruiting, a neo-Nazi movement, and ethnic violence designed and executed to be broadcast on social media, or unleashing algorithms that created eating disorders or motivated suicides. Facebook would just deflect with versions of the same talking point: "What you are seeing is anecdotal, an anomaly. The problem you found is not representative of what Facebook is."

Facebook also loved to remind us that the personalized "world" we saw in our News Feed was heavily determined by our own choices and actions. They claimed our Facebook experience was made up primarily of content from our own friends and family, associates, with whom *you* choose to connect on the platform, the Pages *you* choose to follow, and the Groups *you* choose to join. "Watch where you point that finger," Nick Clegg seemed to be saying in his 2021 editorial "You and the Algorithm: It Takes Two to Tango." Clegg, a savvy former member of the UK Parliament, charmingly pivoted responsibility from the company to users around the world, who had no way of knowing just how much Facebook was manipulating and using them. It was the sort of polished deflection Clegg was paid handsomely to present in order to deflect from the reality that Facebook was progressively filling up your News Feed with content you never asked for, more and more each year, to satisfy their shareholders' insatiable need for ever-increasing profits. The notion that "Facebook is for content from my family and friends" hadn't been true for years—and Facebook knew it.

Call it gaslighting, call it lying—it was intentional. Facebook knew no one on the outside could counter their stories. Furthermore, Facebook knew that only a very few people on the inside knew the company was lying, because only Facebook employees with access to the closed software could see the full picture of what the company was doing. When a user, activist, or government official is gaslit, Facebook steals that person's power to change their circumstances through the truth, and saps the person's or persons' energy to fight back. But once the documents I extracted flowed into public view via an unprecedented, orchestrated strategy that relied first on the *Wall Street Journal*, and later built to a consortium of media around the world, some of that veil of deceit was lifted. Hundreds if not thousands of activists around the world read through "The Facebook Files" and saw years of their work suddenly validated. The public had the proof from Facebook itself that Facebook, just like the Big Tobacco companies before it, had known the toxic truth of its poison, and still fed it to us.

Armed with tens of thousands of pages of Facebook's own documents, and more specifically, reading the meticulously reported news stories and analysis about what was in them, the public had responded dramatically. In the six months after the contents of the Facebook Files had been made public, Facebook's trillion-dollar valuation had plummeted by almost 50 percent and would continue to slide to as much as a 75 percent loss, including the largest single-day loss of corporate value for any publicly traded US company in history. Users in the United States and Europe were fleeing Facebook's platforms. Proposed oversight bills that had languished for years in the maze of government bureaucracies of Europe and the United

States were now zooming toward enactment. Class-action lawyers were circling, demanding justice for the grieving parents who had watched their children suffer and sometimes even die. Facebook could no longer hide from the truth or from the demands of the public to change. We had collectively learned we no longer had to tolerate living in a world defined by Facebook. The era of "Just trust us" was over.

While the public's hunger to learn more continued to intensify, and while people began to experience some catharsis at no longer having to live in a gaslit confusion and began to reconcile their lived experience with Facebook's lies, my own lived experience had transformed. I had morphed from a nearly invisible data scientist and product manager to a surreal new life as the Facebook whistleblower. It felt as if the world viewed me not as me, not so much as a person, but rather as a symbol. As a name, a thing in the news. All of a sudden, I was doing world tours and press junkets. I was sitting in meetings with threat researchers discussing how trolls on the dark web were cyberstalking my mom in the middle of Iowa, dissecting her social media history and plotting potential actions against her and me. Even months later, I would still have days when multiple journalists would ask me, "Are you holding up okay? How has your life changed?"

Originally I had zero intention of revealing my identity. From the beginning, I had two basic goals: I wanted to be able to sleep at night, free of the burden of carrying secrets I earnestly thought endangered the lives of tens of millions of people, and I wanted to be able to drive change from the background. But quickly I learned that I didn't know very much about what it really means to be a whistleblower—and about what it really means to be myself.

I had enlisted the help of a nonprofit legal aid organization that had supported a diverse range of government and corporate whistleblowers for years. They guided me on how to legally disclose information to the SEC and Congress, and how Congress can share information with reporters in a protected manner.

To ensure that the first public interpretation of the documents was as clear and accurate as possible, I also worked closely with Jeff Horwitz, a journalist for the *Wall Street Journal*. We had met in person for the first time on a hiking trail in the Oakland hills a little over a year earlier. By then, I had vetted Jeff carefully and had become confident that I could work anonymously with him to get the truth out. Jeff joked he was the most knowledgeable person about Facebook who hadn't signed a Facebook Non-disclosure Agreement. That's probably accurate. I thought he had the right focus. He was one of the most dogged journalists detailing Facebook's deadly impacts. I knew Jeff could help translate the complex reality of Facebook into a clear picture for the public. I thought he could be the public voice, and I could stay in the shadows.

As publication day for the first *Wall Street Journal* article approached, my lawyers' conversations with me shifted to questions around what I wanted to do once the information was in the public arena. Their guidance was simple and stark: I could do whatever I wanted, but the way they saw it, there was only one plausible path forward: I would have to come out and live openly in my truth and defend that truth, in order to live my life.

My chief advisor and lawyer was Andrew Bakaj, a former CIA officer who himself had once been one of the legal nonprofit's clients. Before advising me, Andrew had advised the person who was probably the group's most famous client: the anonymous whistleblower who first reported the "perfect" phone call between President Donald Trump and Ukrainian President Volodymyr Zelensky to the Intelligence Community Inspector General. That call, of course, was the basis for the first impeachment of Donald Trump.

One of the most notable (and material) details of that impeachment was that the whistleblower remained anonymous. Major news outlets thought they knew who the person was, but refused to publish their name. This was not an accident. Andrew clearly laid out for me what it took to keep the most critical name of the impeachment a secret. Every day for weeks, he spent hours calling media outlets, saying, "If you reveal who you think the whistleblower's identity is and that person is harmed in any way, we will make sure everyone knows the blood is on your hands." Horrifyingly, often the person who the media thought was the whistleblower was wrong — keeping your identity secret can put others in harm's way.

Andrew provided me with a vivid portrait of what life behind the anonymous curtain was like. I would likely spend years wondering what would happen if my identity were revealed. As the impact of my disclosures grew, he said, I should assume "the Facebook whistleblower" would draw a swarm of journalists hunting for the person who had revealed the truth about the social media company that acted as the internet itself for over a billion people and touched the lives of 3.2 billion every month. Media and operatives would want to out me, ostensibly in order to assess my "true" motives.

That's precisely what happened with the Ukrainian whistleblower. Media and other investigators dug for that person's identity. Politicians weaponized the whistleblower's anonymity, teasing speculations about the person's identity in speeches and even trying to out the person with a written question that John Roberts, chief justice of the United States Supreme Court, refused to read at the impeachment. People were obsessed with unveiling that whistleblower. Andrew advised me that if I remained anonymous I should expect the same, for all kinds of reasons.

On the surface, the mystery of my identity lent itself to an archetypal human story: some David standing up to a menacing and seemingly invincible Goliath. Even though we expected most people would view my actions positively, some would not. And on the surface, there were not the same risks to my life that swirled around the Ukrainian whistleblower. My lawyers would not be able to tell media outlets that if they found me out and published my name I could be physically harmed.

In light of all this, I considered another factor. I suspected that the Ukrainian whistleblower had assessed (quite reasonably) that the existence of a transcript of the conversation where President Trump asked the Ukrainian president for a favor in exchange for aid for the country's defense was clear enough that the document itself was all that would be needed for the public to be informed and the case to be made. However, I imagined that the whistleblower perhaps did not anticipate how their anonymity, their absence from the process, would become a means to distract from and undermine the substance of the disclosure.

When the time came for a verdict to be rendered in the impeachment, the defense seized on the absence of the whistleblower, and focused their energies on discrediting and casting doubt on the motives of the person who chose not to be present instead of discussing the implications of the United States denying military aid to Ukraine to defend against a potential Russian invasion. Which, of course, happened, and would now be a critical topic of tonight's State of the Union. Without a face and voice to counter the fictions used to undercut the truth, the weight of the evidence was muddied and eroded.

It seemed to me that my disclosures were radically more complex than a transcript of a single phone call. Without a voice from inside Facebook that could clearly and definitively distill exactly what those documents reveal, without a voice from inside Facebook who could authoritatively connect the company's pernicious algorithms and lies to its corporate culture—just as with that first Trump impeachment—those responsible might not be held accountable. Facebook's gaslighting and lies might still prevail.

Within the 22,000 pages was deep context on how the company's products, like Facebook and Instagram, were designed and functioned; how the employees of Facebook believed they should operate. This was not information that just any person on the outside could intuitively understand, no matter how smart, no matter how educated they were. To become an expert in similarly complex fields, one can get a master's degree or a PhD. But when it comes to the dynamics of Facebook's social-networking recommendation systems and their consequences, there is no academic preparation that could make you adequately informed to independently parse all the corners and nuances of the disclosures.

It seemed unlikely that many on the outside could understand how Facebook's unique culture birthed their unique closed-system software. The only path to deeply understanding these systems is by working at one of a handful of large tech companies in specialist roles. As a result, when I came forward, there were maybe three hundred or four hundred people in the entire world who understood deeply enough how these systems worked to understand why these documents were so damning and who would be able to see clearly that the fundamental threats they documented presented existential threats to humanity.

With Jeff's stories about to publish, I could not wait any longer to decide whether I wanted to be in the public eye. I could no longer cling to the fairy tale I had told myself about being an advisor behind the scenes, trying to split the difference between impact and safety. I could choose to avert my eyes from the fact that Facebook ultimately would know I was the whistleblower by working backward from the documents that were released and accept that at any moment they would be the ones to decide how to introduce me to the world, *their* way. I listened to my advisors and their hard-earned, real-world whistleblower experience.

To tease apart how society and Facebook became entangled in our dystopian dance, what was needed was someone who came from within the company and who was privy to the culture, internal machinations, and interacting demands that the different departments imposed on each other. Someone was needed who could provide the context and connective tissue to understand why so many smart, kind, conscientious people could render a product with such horrific and world-rocking consequences. Perhaps most importantly, someone was needed to warn that opaque companies like Facebook posed unprecedented oversight and governance challenges: that Facebook would be only the first, but not the last, opaque company to wield such extensive damage on the world.

After considering all of this, I decided that someone would be me.

I came forward because I wanted the world to veer from the deadly course Facebook had placed us on. I believed the only way for that to happen was for me to provide briefings explaining what was in the documents and answering the questions they would generate. I also understood how absurd my allegations sounded. If someone said to you, "Did you know that an app on your phone chooses what issues you get to vote on before you enter the voting booth?" you'd roll your eyes. You might chuckle and say to yourself, "Nice conspiracy theory you got there." Maybe you'd say it out loud if you were less polite.

There's almost no way you'd believe that it wasn't a single political party but many groups both on the right and on the left that raised those concerns to Facebook. Each would complain that the influence of Facebook's product and algorithms on the public forced parties and candidates to embrace extreme issues they knew the majority of their constituents didn't like or want, but they felt they had to because it was what the algorithm amplified. It was beyond belief that something that seemed like science fiction could be true. But it was true. I knew all of it was true. I saw it. I was present. I was there. And no one at Facebook could gaslight that away. All of that was literally what the tens of thousands of pages of documents said, if you knew how to read them.

Facebook is a for-profit company that had the opportunity to operate in the dark, and when offered the chance to cut a few corners, it cut them all. After all, Facebook *created* the corners inside its closed software. And if the public doesn't know that a few corners or more have been cut, were they ever really cut? The company had started out as a benign way for Ivy League college students to stay up to date with their friends, and it had continued to exploit that perception to mask a slow evolution into something new. It was no longer a human-scale network of our family and friends, but rather a hyper-amplification machine powered by multimillion-person groups and algorithms that gave the most attention and prominence to the most extreme ideas.

No single person sat down and *intended* to drive the company toward harmful ends. Facebook was a company that fetishized consensus and a mythical vision of itself where all were equal (besides Mark Zuckerberg, the CEO). When I joined in 2019, its Menlo Park, California, office held the record for the largest open-floor-plan office in the world, clocking in at a quarter of a mile long. For years, when pressed before Congress, Facebook executives always refused to disclose who was responsible for which decision — committees made decisions, they insisted, there's no single responsible person. But without individual responsibility, there is decreased motivation to stand up and say, "This is unacceptable," or even pause and ask, "Should we be doing this?" Ultimately Facebook had formed a culture that did not value personal accountability. How and why did that culture take shape? How did it function day by day? I could explain that, too. And how those cultural dots connected to the code behind the algorithms.

By the time I arrived at Facebook in 2019, people had been aware for at least a year that the company's decision to shift from just trying to keep you using its products for as long as possible to trying to provoke a reaction from you had driven a surge in extreme content. Facebook had made this shift in late 2017 into early 2018 in response to a slow but troubling decrease in the amount of content being produced on the platform. The company had run many different "producer-side" experiments on people who posted content on Facebook and found that the only intervention that increased the amount of content produced was giving creators more small social rewards. In other words, the more people who like, comment on, and reshare your content, the more likely you are to produce more content for Facebook.

Most people think about social media companies only from the consumption side. As in, I go to Facebook, to Twitter, to TikTok to consume content. This is a reasonable association, because the vast majority of actions and time the average person spends on social media is spent on consumption. Social media companies, however, think of themselves as "two-sided" marketplaces that connect people who want to create with people who want to consume, just as a marketplace might connect people who want to sell with people who want to buy. You can't have a buyer without a seller. You can't view content unless someone first produces it.

Under our current corporate law and policy, Facebook also has a duty to its shareholders to generate ever-increasing profits. There are a relatively limited number of paths to accomplish this. They can create or buy entirely new products; they can recruit more users to their current products; they can drive more money per ad from the users they have; or they can get those users to consume their products more extensively, because consuming more content leads to viewing and clicking on more ads. All of these mechanisms allow the company to profit by selling ads to advertisers that are in aggregate worth ever more money. And all of it depends on user habits — natural habits or created habits.

By 2019, Facebook's antitrust concerns had frozen the first path of expansion. They were not going to be allowed to merge with any more social media companies. Some people were even talking about breaking Instagram and WhatsApp off of the core "Facebook Blue" business to spur competition. The second path also showed a lack of promise. The vast majority of internet users had already signed up for Facebook's products. Facebook had been investing extensively in subsidizing people to use its products in ever more economically fragile corners of the world (and closing off the opportunity for the free and open internet to form), but those users individually drove little revenue, and thus the third path was also closed.

That left one remaining path, which meant getting people to consume more content. In 2018, to arrest the decline in content production, Facebook shifted how it ranked content on its News Feed to prioritize content that produced more likes, comments, and reshares. Facebook is constantly and subtly training you on what kind of content belongs on the site, intentionally or unintentionally. While influencers and other power users like publishers carefully study what gets distributed on social media platforms and consciously adapt to produce content that is similar to the most distributed items, most individuals subconsciously also alter what they create for social media based on what they see in their feeds. What you see in your personal feed subconsciously becomes "This is what Facebook is for." In 2018, as Facebook began to give more distribution to content that provoked a reaction, individuals across the world began to see certain kinds of content on Facebook more often, even if they weren't aware that was what was happening.

By December of 2019, its data scientists were pointing out that Facebook had created a feedback loop that could not differentiate between positive and negative reactions. You might drop an angry face on a post or write a comment, saying that you hated the article or that the article itself was misinformation, but the algorithm just took it as a sign to show more similar content to you and to others, because you had engaged with it. Publishers began to see that the angrier the comment thread under a link, the more likely you were to click back to the source website. It always annoys me when people give Facebook a pass because news outlets or websites run sensationalized news. The majority of media outlets need to be profitable to continue to exist, and they tune what they create (just like political campaigns) based on what the platforms share with consumers.

While this feedback loop had contributed to souring political discourse in the United States and western Europe, in the most fragile places in the world it had contributed to the deaths of tens of thousands of people by pouring social media gasoline on communities already struggling with the sparks of ethnic tensions and historical grievances. Facebook had faced the first large-scale communal violence incident in 2017 in what Amnesty International described as a "social atrocity" in Myanmar, a country in Southeast Asia. The country's own military had established a network of tens of thousands of accounts, pages, and groups run by a staff of seven hundred military personnel to distribute and amplify propaganda targeting the Rohingya people.

The *New York Times* reported that Myanmar had years before sent large groups of officers to Russia to study psychological warfare, hacking, and other computer skills. Even then, Russia had established itself as the world leader in social-media-driven cyberwarfare. Russia's cyberops were a large part of its attack on Ukraine. Anyone following the news was aware of that. What I knew, as I took my seat in the guest box for the State of the Union, was that Facebook had failed time and time again to address its seismic role in Russia's cyber operations, or rather, what I knew was that Facebook had *chosen* to ignore what it was facilitating for (and with) Russia, for example.

This investment that Myanmar had made in training and establishing a broad social media network to parrot propaganda sprang into motion in 2017. The military used it to distribute lurid photos, false news, and inflammatory posts, often aimed at Myanmar's Muslims. Critics found it difficult to oppose the fake allegations because troll accounts run by the military ganged up on anyone attempting to defuse the conflict and added fuel to arguments among commenters. The military trolls took a page from what would become the standard ethnic-violence playbook across the world: They would post out-of-context or otherwise fake photos of corpses that they said were evidence of Rohingya-initiated massacres.

I came forward because already, in 2021, the second wave of large-scale Facebook-fueled violence had taken shape, this time in Ethiopia, loudly echoing what had occurred in Myanmar only a few years earlier. I deeply believed and believe today that the choices Facebook had made about its products and their rollout around the world would endanger the lives of tens of millions of people over the next twenty years. I wanted to be certain that the people with the ability to intervene understood the depth of this international, destabilizing, and worsening crisis, and the only way I could ensure that, I decided, was by sitting with government officials until I felt confident they understood exactly what the stakes were.

When we reached the First Lady's box I was seated next to Valerie Biden, the president's sister and long-time campaign manager. One of the other guests was Danielle Robinson, the widow of US Army Sergeant First Class Heath Robinson. After surviving deployments to war zones in Kosovo and Iraq, he died of a rare lung cancer caused by extended exposure to US military burn pits. In the wake of her husband's death, Danielle had dedicated herself to trying to get support for the families of veterans who are sick or have died because of the burn pits created and managed by the military-industrial-complex contractor KBR, then a wholly owned subsidiary of Halliburton, a "strategy" that prioritized short term-profits over soldiers' lives. Ukraine's ambassador to the United States, Oksana Markarova, was seated a few seats away. As we entered the box, she handed each of us a small Ukrainian flag.

Being surrounded by people who had sacrificed and lost so much, and yet who not only endured unbroken but emerged resolved to have hope and effect change, was overwhelming. I felt as I often have throughout my life: like I didn't belong. I had done my very best that evening to present myself in such a way that showed my respect for this solemn historical moment and for such honorable and distinguished guests. Just figuring out what to wear and how to appear had caused me some stress.

My mother was a trailblazer at the University of Iowa. I was the first baby born to a female professor in her department. My existence symbolized her commitment to having a family and the hardship she had to overcome to have me in her life. As an assistant professor, she was told repeatedly she would be throwing away her tenure track career if she were to have a baby. Every day she wore plain dresses to her work as a professor in the Biochemistry Department, in part as an expression of her disinterest in clothing and in part as a preference for efficiency over display. She did not color her hair when she started to go gray because, she said, she was often the youngest person in her committee meetings already.

It had taken me until I was twenty-five and had arrived at Harvard Business School (HBS) to realize how much I had missed out on. I hadn't learned (from my mother or from my peers in tech) how to access the traditionally feminine parts of our culture, or of myself. Prior to that, my first job out of college was at Google, with very few women more senior than me to draw upon as role models. The few women I had known at Google (less than 10 percent of the search quality team at the time were women) had largely followed the same path as my mother, attempting to defeminize themselves as protective camouflage. There was only one prominent exception to the rule: the head of the management rotation program I was in, the vice-president of Search, Marissa Mayer. Not until I encountered at HBS women from the full spectrum of industries that make up the world economy was I exposed to women who were both powerful and — because they had been trained in female-dominant industries — saw no conflict between being competent, powerful, and also beautiful.

Before coming forward, I rarely wore makeup — maybe only a handful of occasions in the decades since my teens. Thank goodness the same makeup artist *60 Minutes* had provided for my interview in mid-September 2021 was available to help again this evening. I wore a dress I had picked up at Nordstrom Rack the day before. The dress was teal and, as I had been counseled, "knee-length and not too busy." Thanks to the advice of a trusted friend and colleague, I wore a scarf, both to complement the dress and to provide a bit of warmth. At the last minute, just before we left for the pre-State of the Union dinner at the White House, a friend had secured and delivered to me a pin of the Ukrainian flag.

From our vantage point perched above them in the box, I could see all of the members of Congress who were present on the floor of the House. There was the thinnest facade of unity. The reality was that the United States in 2022 was dramatically polarized. I knew firsthand that the division and tribalism had in no small part been relentlessly charged by the engagement-optimizing algorithms across Facebook. After two years of a global pandemic and more than a decade of exposure to a social media information ecosystem that rewards and promotes extreme content, the State of the Union and the world were bitterly divided.

Given how divisive and politicized the response was to former president Trump's election in 2016, the United States was in a vulnerable position in which many on the right believed claims of Russian interference were overstated sour grapes from people who opposed Trump. Now, six years later, many on the right believed that the 2020 election had been stolen, thanks to our warped information ecosystem and Russia and Ukraine waging extensive disinformation campaigns and cyberwarfare against each other along with the missiles and planes they launched at each other.

We live in a world where the weaponization of social media is considered by militaries to be a vital aspect of war, yet many in the United States and abroad accepted Facebook's framing of the problems of social media and the available solutions. Facebook's foremost PR victory of the previous decade was tricking us into believing a false forced choice between "freedom" and "safety," that we had to choose preserving "freedom of speech" over "censorship." Many people, including many people in that chamber below me, did not want to discuss fixing Facebook's problems, because they were quite reasonably opposed to censorship. Facebook had convinced us that those were the only two choices, when in fact the company had thousands of pages documenting a world of alternatives.

President Biden swept into the chamber just as I had watched previous presidents do in countless other televised State of the Unions. Only now I was...there...*here*. He began his speech, as expected, with an outline of the situation in Ukraine and the need for the free countries of the world to stand with those who did not ask to be invaded. We all waved our Ukrainian flags.

No one tells you in advance when in this speech your name will be mentioned or how you will be introduced. When the president invoked my name, I was completely caught off guard. "Frances Haugen, who is here tonight with us, has shown we must hold social media platforms accountable for the national experiment they're conducting on our children for profit. Folks—thank you. Thank you for the courage you showed." Just like that, before I realized what had happened, I stood and then sat down, dazed. My head was spinning.

It had not been a smooth, linear path to this night, to this balcony, to this reckoning between the public and one of the world's largest corporations. My journey was not that of a mythical hero, but of a small and different girl who persisted over and over again in small steps that added up over a long period of time. Of a teenager and young adult who started by refusing to let others tell her not to exist or to get back into the box they thought she belonged in. It had been a journey of learning that I could make choices, make decisions about my life, and ultimately, that any one person and any one decision can have enormous power. All of us have more power than we realize, even if it might scare us to accept that.

We live in a world where a sense of fatalism easily creeps into our lives—a sense that the problems we face are insurmountable. That we are each too small to meaningfully impact anything. I want to be clear: When you feel fatalism, it's a sign that someone is trying to steal your power. It isn't always easy to see that. Hundreds of thousands of employees had passed through Facebook's doors before I had, and had not acted. Many had burned out and left the company. Many others had stayed, worked very hard, and accommodated themselves to the world-defining platform they were helping to create.

Imagine if we all realized the power of one. What world could we build together if more people woke up to their own power?

CHAPTER 2

When I Was Young in Iowa

Life is a matter of choices, and every choice you make makes you.
—John C. Maxwell, *Beyond Talent*

I was sixteen years old, standing beside an open casket. I felt hollow. I didn't know what I was supposed to do as I looked down at my co-captain and longtime friend. The casket was open, though it shouldn't have been. The Tina I knew had worn her hair curled by her ears; when she smiled dimples took over her slender face. She wore glasses and sometimes when she would throw me an impish glance it was like her eyes were giggling. The girl in the coffin had on Tina's glasses, she had Tina's hair, but her face was badly swollen; her mouth, eyes, all of her facial features were askew. This wasn't what the Tina I loved had looked like. I suppose that whoever chose an open casket for the viewing believed it was for the best, thinking that if we could see her one last time it would somehow help us grieve, help us say goodbye. All these years later, I remember vividly the horrific circumstances of her death: that Memorial Day weekend on I-80, the overcast skies, the slick roads. Funerals are supposed to be about letting go, yet the morticians clearly had to work to make Tina even remotely presentable, and this person was a stranger to me. This Tina in the open casket did not help me connect so that I could say goodbye and, most of all, so that I could say thank you. Many years on, I can't help being aware of how Tina's presence and then her absence forever altered my life.

Tina Wang had been one of my closest friends since junior high school. I don't recall exactly how we met; it probably was in the cafeteria. Tina and a handful of other Chinese American students tended to sit together, and their lunch table soon became my lunch table. It might have been that we started talking after one of the classes we took together, as we entered the chaos of kids in the halls during change of classrooms. I wish I could remember, but I don't. I do remember this: Tina accepted me, she befriended me, and she allowed me to befriend her when few others did.

When you're a kid, you cannot really understand the potentially seismic effect of the small choices you make. No matter how smart you are as an elementary school child, there's simply no way you can begin to conceive how a choice you make—or a choice someone makes for you—in a given moment can forever shape your life, can profoundly impact who and what you will become, what opportunities you will have or not have, or what challenges you will face and how you will face them. While I was an intellectually gifted kid, I didn't appreciate such a concept as I embarked on junior high school.

Most people are forced to endure only one first day of junior high, but because I insisted on altering fate, I had two first days: the first day of seventh grade, and then the first day of the second trimester of eighth grade. When I was in elementary school, Horn Elementary had many students who were children of professors at the University of Iowa, located a few miles away in Iowa City. As a result, the teachers at Horn were accustomed to precocious children coming through on a regular basis. It was not uncommon for a child to be taking calculus in sixth grade, or for a first-grade student to be reading at the college level. There were kindergartners (like me) who had college-level vocabularies. The Horn faculty was extremely good at providing exceptional students with the lesson plans their brains needed, while never losing sight of the fact that we were in fact children.

My mom intervened at a critical moment for me in elementary school. At the start of third grade I had begun coming home from school crying because I was so bored in math class, and my mother gave the administration a simple directive: "My child has special needs and needs reasonable accommodations." This magic phrase from the Americans with Disabilities Act unlocked the full flexibility of Horn Elementary's curriculum, which relied heavily on self-paced projects that allowed kids to go as deep as they wanted to go rather than separate them into accelerated and standard groups. It was a system that valued keeping children integrated into their peer communities. As the end of sixth grade approached, Jan Bohnsack, my gifted teacher, began to worry about my transition into the general population of the junior high school. Given that I had become accustomed to Horn's flexibility and the ability to race ahead, she was concerned I was about to hit a brick wall when I would be forced to slow down and wait.

She called a meeting of my parents, the administrators, and me in which she strongly advocated that I skip seventh grade. Instead of trusting her judgment, I suggested a compromise: I would start the first trimester in seventh grade and then do the next two trimesters in the eighth. This ostensibly small choice was the first of so many I would make that would radically affect the course of my life.

The principal of the junior high was a strong believer in the divide-and-conquer school of governance. He pitted the eighth graders against the seventh graders (the "sevies") to maintain order. By transitioning from seventh to eighth grade more or less in the middle of the school year, I had set myself up both as a defector from the seventh graders while not being a "real" eighth grader. Junior high is hard enough for tweens trying to fit in, to make friends in a new environment and community. Now I had made that already challenging period even more challenging.

Northwest Junior High was one of those sprawling, one-story, academic buildings from the 1970s that gave off the vibe of a low-security prison, and the administration knew it. This is where I met Tina and through Tina the group of Chinese American kids who were all close-knit friends. They didn't make fun of me for being good at school and for caring about my classes. They did, too. In the 1990s it was clear that Iowa was economically struggling and its future didn't look brighter. This circle all had the same goal that I had independently formed — even then, we all felt we needed to be able to get out of Iowa. Getting into a good college was a vital step in that plan. They accepted me; these were now my friends.

It was in junior high that I realized I was extremely skilled at math. I was selected for the state MathCounts math competition team for my school, alongside John Hegeman, whom I would see again decades later at Facebook, where he headed the team that chose how to prioritize content in the Facebook News Feed while I worked on Civic Misinformation. Many math competitions are divided into two styles of competition: a sprint round with thirty problems to solve in a set amount of time, and a target round with only six problems to solve, but the problems are significantly harder. For the latter, I came in second of all junior high students in the state, despite chronologically being a seventh grader.

The high school I transitioned to, Iowa City West High School, was a slightly older but much more inviting building than my junior high. It had big windows and high ceilings, and — Iowa being Iowa — was nestled between a cornfield and a Mormon church. A block away from the high school was Mormon Trek Boulevard, commemorating the time when the Latter-Day Saints, having fled to Utah after facing a pogrom in Missouri and Illinois, summoned European Mormons in Britain and Scandinavia to join them. Thousands sailed across the ocean and then rode the train as far west as they could go (which happened to be Iowa City at the time). From there they walked and rode wagons thirteen hundred miles across the prairie to Salt Lake City, many of them so impoverished that all they brought with them were three-by-four-foot handcarts. Brigham Young led the Mormons west to establish the Salt Lake colony in 1847 to escape being killed. In an oft-forgotten bit of history, Missouri's Mormon Extermination Order issued in 1838 stated that "the Mormons must be treated as enemies, and must be exterminated or driven from the State." Shockingly, this order wasn't formally rescinded until 1976.

I regularly rode my bike past a large bronze statue that commemorated that forced migration. When I talk about the dangers of Facebook and ethnic violence, it doesn't feel abstract to me. I knew from a very young age that even in America we have slaughtered religious minorities because we were divided among ourselves and scared of them.

I entered high school feeling very alone. I suppose like many teenagers on the first day I didn't know who I was or what I wanted. But unlike many angst-ridden freshmen, I started high school with virtually no close friends, with the exception of Tina.

At Iowa City West, I was fortunate that I found teachers and organizations that gave me a framework to begin to recognize and become myself, teachers like Mrs. Muhly. The first time I began to appreciate my self-worth was because of her. She and I had first encountered one another right before I started high school. Mrs. Muhly spoke with a thick Brooklyn accent that stood out against the neutral midwestern accents of Iowa. In high school, she was the first person to tell me I had the potential to accomplish great things, but only if I didn't sabotage myself.

Both for the intellectual challenge and to better position myself for college, by the time I entered high school I already knew I wanted to take math classes at the University of Iowa during my senior year. In order to do that I would need to reaccelerate from the relative delay introduced by my skipping a grade. This didn't seem like a big deal to me. I had taken geometry during a summer camp in elementary school, and I knew from the SATs I had taken a few years before as a screening test for a summer gifted program that I was already at the ninety-fifth percentile for high school seniors in geometry. The way I saw it, taking Geometry Honors concurrent with Algebra II Honors was eminently doable. To do so, I would need the sign-off of Mrs. Muhly, the head of the Math Department.

Algebra II Honors was Mrs. Muhly's prize class. These were the students who won the math competitions and filled her calculus sections a few years later. No, Mrs. Muhly said, I most definitely could not double up those classes. She was of the mind that a mastery of geometry was required in order to excel in Algebra II Honors. She wasn't impressed by the summer course I'd completed. If I insisted on taking both, she said, I could take regular Algebra II and Geometry Honors. She explained that she was valuing my long-term happiness over my short-term desires. She wanted to maximize my chances of success rather than letting me get in over my head.

There is much discussion in the education community about the value of "tracking," the dividing and grouping of students according to their ability based on test scores and maturity. Advocates of tracking say that when students in a given classroom are learning at different speeds, slower learners get left behind while more advanced students get bored. In regular Algebra II, for the first time I experienced the flip side of that argument: lower-tracked classes aren't given adequate resources, and students rise or sink to the level of expectations. If you teach with less rigor (or less content because you go slower), such tracking further widens the educational gap between the students.

Until then, I had always been in math classes with advanced students who all expected to go to selective colleges. Now, sitting in regular Algebra II, I could see the impact of those different standards. Even the teacher was a controlled variable. The same man taught me Geometry and Algebra II in back-to-back periods, making the difference in how he treated the two groups of students even more apparent. He permitted the algebra students to be less disciplined, so the classroom was noisy. All in all, there was much less of the devoted attention and energy my previous peer group had demonstrated. The homework standards were lighter and the depth of our teacher's explanations not nearly as thorough.

I wasn't in regular Algebra II very long. For all our math classes, the district made every student take a test at the beginning of the course and again at the end to assess baseline knowledge and subject-matter growth. At the end of the first week, after we received our pretest scores, Mrs. Muhly pulled me aside. She told me she had made a mistake — I had gotten the highest pretest score in the school — and she apologized for not letting me take Algebra II Honors.

Almost immediately Mrs. Muhly had me in weekend math competitions, where I regularly ended up ranking and sometimes winning for my grade. But to her surprise, come Monday, when it came time for me to take regular math tests in Algebra II Honors, I would get B-pluses. Mrs. Muhly called me to her desk after class one day and asked me to explain what seemed to her like a consistent discrepancy. I told her that I had been doing my homework in her class for the following class, and I rushed through her exams so that I would have time to finish my homework before the next period began and that homework was due. Her response was swift. She informed me that until I could demonstrate that I was responsible and understood I needed to get good grades in math to get into college, I was going to take my tests after school in the hallway. After a couple of rounds of afternoon test taking, I straightened up and got with the program.

Mrs. Muhly gave me a new perspective on what I expected of myself and on how I cared for myself. I regarded her apologizing to me as indicative of her integrity, of how seriously she took her job and of her respect for me, even though I was just fourteen. I felt seen and valued.

Tina Wang was the one responsible for encouraging me to make the most important decision I made in high school, which would turn out to be one of the most influential choices of my life: to join the debate team. The experience and skills I developed and the knowledge I obtained in debate profoundly informed the work I would do as the Facebook whistleblower.

In our freshman year, she encouraged me to join when she did, saying, "It's good for getting into college. Try it at least." The team was small, maybe five or six varsity debaters and another six of us first-year novices. We joined a program that was passing through a lull. In the past, the team had regularly produced national-level tournament winners. By the time I joined, that success was a memory, leaving us one of the few high schools in the country in which the debate program had the same budget as the football team, despite having only a handful of debaters.

Debate is one of those extracurricular activities that can easily absorb a vast amount of your time, and I was more than happy to be absorbed into it. Our designated room was one of the few places in my high school, along with the drama club in the school theater and the budding journalists in the school newspaper office, where you could find people regularly hanging out and working into the evening. The debate room became a refuge, a place where I felt a sense of belonging beginning the very first day, when they gave me my own cubby and labeled it with my name.

It was on the debate team that I first recognized my significant interpersonal deficits. I couldn't pick up on sarcasm and wouldn't get the humor of my teammates' jokes. Our coach (until the end of my sophomore year) fondly referred to me as the "absent-minded professors' child." Communication at home between my parents, brother, and myself was simpler, matter-of-fact. No raised voices. No jokes. There were no double meanings. Conversation served a purpose and was direct, earnest. And because I had been so isolated during my brief sojourn through junior high, I couldn't tell when people were being serious.

That same group of Chinese American friends I had had in junior high was now the nucleus of the novice debate team. Along with Tina, Longwei and Xiang also joined. Because of the team's robust budget-to-debater ratio, we traveled together to many weekend tournaments across the Midwest, and later across the country. I loved the thrill of the debate rounds, the feeling that I was part of some secret society of smart, funny, well-read outcasts. I was riveted watching senior debaters in elimination rounds joust with each other, experts at their craft. I find it amusing that I still slightly swoon over loosened ties worn with collared shirts, because that's how the senior debaters looked when they debated in finals. They no longer needed to look put together, they were taking care of business; they were that good.

I wasn't expecting the debate team to be a place of drama, but where there is life there will be some unavoidably trying times. One fall day during my freshman year, I came into the debate room and learned that the girlfriend of one of our senior debaters had died over the weekend of a heart attack. She was a near-daily fixture hanging out in the debate room. She was as petite as a sparrow, yet radiated such outsized energy I often marveled at how such a tiny human could emanate that much enthusiasm. I had not been aware of it, but she struggled with bulimia. Her ritual of vomiting up her food before she could absorb the calories had altered the balance of electrolytes in her body to the point of causing fatal heart failure.

Decades later, when I would discuss the dangers of Instagram with experts in teenage mental health, I would realize, even twenty-three years after this death from complications of bulimia, that the loss still hurt me. Today, when I talk to parents and they say things like, "I know my kid isn't struggling with social media," I always *want* to believe it's true. Yet I know the role Instagram plays in promoting eating disorders. Facebook's own documents discuss how Facebook isn't just bad for kids' mental health, it's significantly worse than other forms of social media because of what it focuses on and how the product is designed. TikTok is about performance, humor, and doing things with friends. Snapchat is about faces and augmented reality. Reddit is at least vaguely about ideas. But Instagram is about social comparison and bodies.

It's easy to trivialize eating disorders as just "skinny girls or women," but studies out of UNC Chapel Hill, Baylor, and Harvard Medical School have suggested serious obstetric and gynecological problems associated with disordered eating; even girls who "grow out of" such eating disorders are doing potentially irreparable harm to themselves. Some of these girls one day may want to start a family, only to find they are unable to conceive because of the damage they've done to themselves in order to look like the women they follow on Instagram. Because of Instagram's influence on culture, there will be women walking around this earth in sixty years, stepping tentatively through their lives out of fear that their bones might shatter.

I always respond the same way to parents who say their kids are okay. I tell them I believe them, and I follow up with a question: "Are you equally confident every other child your son or daughter knows has a healthy relationship with social media?" Our children are intimately interconnected. If I can still feel pain more than twenty years later from witnessing the harms of an eating disorder in high school, no child is safe until all children are safe.

Almost immediately after joining the debate team I endured emotional abuse, becoming a target of the rage of our coach, who had obvious anger management issues. Typically, when the coach became unhinged, he would start with my debate partner, Olivia. She was a year ahead of me, and when it came to dealing with these rants, she was light-years ahead of me in experience. Olivia's father had similar anger issues; he would unload on her at home. Dealing with her father, traumatic as it was, had steeled Olivia's nerves. Whenever the coach would tee off on her, she would simply stare at him expressionlessly, giving him no reaction. Then he would turn on me.

Like many teenagers, I did not have a strong sense of self. I quietly grappled with insecurity. I had become dependent on external sources of validation, especially from my teachers. It was how I defined myself. A teacher's approval sent my confidence soaring; a teacher's critique crushed me. When our coach would rant at me about the ways I was apparently failing, it was devastating.

We developed a routine. The day after one of his outbursts, he would pull me out of class and apologize to me in the empty hallway, a ritual I knew wouldn't deter him the next time. I relied on the debate program for stability in my life. I didn't feel empowered to (or ever considered that I could) go to the administration and report his behavior. I felt trapped,

trapped in the kind of unhealthy dynamic that would repeat often throughout my life until I became more self-aware, until I learned how to set boundaries, and when necessary hold my ground and take no more.

———————

For those of you who were not debate nerds, there are two styles of competitive debate: Policy and Lincoln-Douglas. I credit my participation in both with providing me with the philosophical foundation and ethical groundwork that allowed me to understand both my obligations to act as I did at Facebook and the practical skills to successfully collect and explain my disclosures.

Policy debate is pretty much what it sounds like. You get a topic or topics, and are expected to learn and master the issues in order to argue alongside a partner for or against any policy within the scope of the resolution. When I did Policy, the topics were so broad that basically all competitive debaters on the national circuit would spend weeks during the summer learning about the subject matter. I spent fourteen weeks at debate camp over the course of my high school years. Fittingly, two of the policy topics I researched at debate camp were protecting privacy, where I learned about the ethics and implications of online privacy, and education reform, where I learned about pedagogy and cognitive development.

Debate camp is about as glamorous as you'd imagine. For me it was like an oasis. Each summer I went to one, first at Wake Forest, then the University of Michigan, and finally at the University of Kentucky. I would spend long hours submerged in the library stacks. It was there and then that I fell in love with large institutional libraries, a love that later would lead me to work on search quality for Google Books.

To prep for policy debates, we would have large Rubbermaid bins (lovingly called tubs) full of manila folders, each full of paper documents on a specific topic. A pair of debaters might have four of these tubs. The ability to keep the hundreds of folders in some kind of order was essential to being able to quickly access them and the information therein as you debated. I fully credit the time I spent at debate camp and in Policy debate with preparing me to pull off the massive disclosure of information and documents from Facebook.

(In 2021, as I collected information for the public, I would take photos of each of the 22,000 pages of Facebook documents I planned to share. I would then place those 22,000 photos in appropriate digital folders. Those folders were then grouped into tranches that represented each time I cleared the memory of the small burner phone I used to take the photos. To keep track of the order in which the documents had been acquired [in case history ever cared] and to make it easy to find a specific doc, I put the number of the phone dump in the title of the outer folder, and the number of the specific document within the folder. That way people could refer to any given doc using two sets of two-digit numbers. For example, Folder 15/Doc 08 was an internal memo, "We are responsible for virality." When I was fourteen, I never would have guessed that the organizational skills I was developing in the University of Iowa's library would play a role in history.)

That debating experience taught me more than just how to research and how to organize information. For one debate resolution, I had to learn everything I could about education reform and education theory. How we educate students, particularly in computer science and related subjects, directly contributes to the culture I encountered at Facebook. One of the things too often missing from technical education today — how most computer programmers are educated — is an investment in helping individual students cultivate their own perspective on how their work can and should impact their own lives, and more importantly, how their work influences society and the world at large. Most Americans are not aware that at many universities in Europe, if you're accepted into a computer science degree program to become a software engineer, you are not allowed to take a single philosophy or sociology class, because you were admitted to the computer science program, not the university as a whole.

The United States is not substantially better. Most engineering curriculums here have more core-course requirements than other undergraduate degrees. This leaves students little time to take anything else, thus leaving few opportunities for them to take classes that would encourage them to contemplate what their responsibilities are as computer science engineers wielding their ever more godlike powers. Every class you take that might help you engage in values-based critical thinking decreases your chance to score that coveted job at Google, or whatever the most sought-after employer is of the day, since it offsets a computer science class you would have otherwise taken.

The ability to assess your life and environment and ask, "Is this what I want? Is this a genuinely valuable thing to do?" is a muscle that must be trained and challenged in order to be effective. On the debate team, I had an opportunity to explore some basic but essential questions like, "What do we owe each other?" and "Who am I?" and "What kind of world do I want to live my life in?" Because I coached debate when I was in college, I was forced to examine what I thought I knew and to become extremely clear in communicating it, even to an audience that was starting with zero context in a topic. Although I had no way of comprehending such things at the time, all of these are vital skills for a technology whistleblower.

During my time working on the Civic Misinformation team and witnessing how Facebook operated internally, I came to the conclusion that the company did not have the right to make some of the value trade-offs it was encountering, or ignoring, in isolation. Debate forces us to examine what we value and also to respect the dignity and autonomy of an individual. Fundamental to valuing the dignity and autonomy of an individual is ensuring that the person has adequate information to be able to consent to their interactions. If we intentionally hide or withhold information from people that would change the decisions they make, we are exerting power over them. That is manipulation. That is precisely what I saw Facebook do over and over again. Not just withholding information, but actively denying the truth when people brought up concerns.

When it was occurring to me that Facebook had organizational issues that precluded it from fixing its own problems, one of the first symptoms I noticed was that many of its employees weren't even aware of conflicts of interest. They would just say, "We don't have enough people to do anything more than we're doing today," at a time when Facebook regularly had profit margins of twenty-five to forty percent on $80 billion of revenue. Facebook had the money to hire more people; their scarcity was a choice. Yet they would earnestly express an inability to act without so much as considering that maybe that wasn't good enough. That maybe Facebook wasn't the most objective arbiter of what was "sufficient." Or that implicit in that Facebook thinking was an assumption that the company didn't owe it to vulnerable populations to spend more on safety so Facebook could afford enough people to meet a bare minimum of safety. But I'm getting ahead of myself.

———————

In my junior year at Iowa City West, my debate partner quit, and I moved from being a Policy debater to competing in Lincoln-Douglas debates because it was an individual rather than team-based event. With the move, it seemed a certainty that I would be the "C" debater on our team, after Shalini and Tina. Being the C debater opened some doors and closed others. I would be allowed to travel to local and regional tournaments, but would more likely than not be kept out of national circuit tournaments that limited entries to just two people per school. Iowa City West had hired a new debate coach at the beginning of my junior year, Scott Wunn, who was so effective that just a few years later he would go on to become the head of the National Speech and Debate Association. In his first few months coaching our team, he succeeded in greatly expanding the scope of the debate and speech teams. He had pitched to me that since I wasn't going to be competing as much as I might like during my senior year, I could step up to be the Novice (ninth grade) policy debate coach.

I wasted no time. Shortly thereafter, I was standing on the stage of the Northwest Junior High auditorium, surveying the thirty to forty fidgety thirteen- and fourteen-year-olds who had gathered that spring afternoon for my first stab at running a junior high debate program, an institution that I would discover twenty years later was still operating. I wanted to lead them to victory as the Novice debate coach during my senior year, and that afternoon was the first step in executing that plan. I had borrowed a page from Mr. Wunn's playbook, and had spent a day doing a road show for all the school's English classes. I must have painted a compelling vision of debate because this mass of unbridled energy was now arrayed before me.

A few weeks later the spring trimester was drawing to a close and I had the feeling we were looking forward to a remarkable Novice debate class and an exciting year of competition. By the time we ran our capstone debates you could see the junior high debaters had a camaraderie that would pull them through the following year. Things were looking great.

All that was left of the season was our Memorial Day fundraiser along Interstate 80, where we would give out coffee and cookies in exchange for donations.

The energy at the rest stop buzzed with activity as fifteen or so teenagers chatted while waiting for drivers on their way to the restrooms to pass our tables. The day was overcast and the intermittent rain seemed to have finally stopped. As the co-captains of the team, Tina and I were helping to keep things humming, and because she had a car, she said she was leaving to go check in on our site on the opposite side of the highway. I still remember her hair on that day, stylishly curled on either side and pinned back behind her head, as she purposefully strode away from the shelter where we were set up.

In hindsight, you realize there are moments in your life when you were living in a fragile sliver of time, pregnant with a Before and an After. A Before when everything was okay, wonderful even, in ways you took for granted; and an After, when everything after that abruptly snap-spirals into something entirely unforeseen and awful. This was one of those moments.

The first sign that something was off was when Mr. Wunn walked up to me and told me to gather everyone together. He announced that we were all going back to the school, without explaining why. I remember conjecturing out loud to someone that I wondered if something had happened to Tina. After all, she had never returned like she said she would. Back at school, maybe forty-five minutes later, we were in the debate room when Mr. Wunn informed us that Tina's car had been hit head-on by a semi. She had swerved on the slick surface and then overcompensated, skidding through the grassy median and into the oncoming truck. She died instantly.

People act in irrational ways when faced with trauma. I can remember only a handful of times in my life when I have actually stolen something. I've always been a believer that the punishment for sin is sin, that if you steal, you fear being stolen from. But at that moment I stole. Maybe I stole because something, or rather someone, had been stolen from me. Minutes after Mr. Wunn told us Tina had been killed, I took her school books from her cubby in the debate room. I still have them to this day.

The next two weeks were a blur. I was unable to eat, and I would sit numbly through classes. Teachers spent time with us, unpacking Tina's loss. Each of those AP classes that Tina and I shared was full of people who had been tracked alongside us since junior high, a small cohort of students in the same accelerated classes. As we moved from class to class, her absence seemed more and more unbearable.

It's said that when we recall an experience, we are most likely to remember the first thing and the last thing we encountered. This period of initial mourning was no different. I don't remember many details of that period, but I know that there was a funeral, and people came from many states because they had known Tina through debate. I remember speaking at the service, but I don't remember a word I said. Whatever I said to convey how much Tina meant to me, how much I loved her, I'm sure fell short of what I felt.

Because Tina was gone, I moved up one spot on the debate team to be the B debater, which meant I traveled and competed. Because of my new ranking, the world opened up to me. I couldn't help but think that even in death, Tina was giving me a gift. I spent seventeen weekends on the road competing over the course of senior year, often driving with our coaches as far away as Texas or flying to more distant places like Florida or Massachusetts. Unquestionably, being able to attend and place highly in some of the most competitive debate tournaments in the country improved my chances of getting into college. It occurred to me that this is what Tina should have been doing. If not for Tina, I never would have joined debate. If not for Tina's death would I have gotten into MIT, CalTech, or Olin? I went into the next phase of my life carrying a self-imposed burden that I had taken my friend's spot in the world and needed to live up to Tina's gift.

CHAPTER 3

Olin: My First Start-up

The direction in which education starts a person
will determine his future life.

— Plato, *The Republic*

Throughout high school my dream was to attend MIT. It was a classmate, Jay Gantz, who pushed me to consider a start-up institution in Needham, Massachusetts, the Franklin W. Olin College of Engineering.

"What do you mean you're not applying to Olin?" he had asked, shocked that it had fallen off my list at some point during the fall of my senior year.

The truth was, I told Jay, I didn't think I would get into Olin. Only thirty-two students were going to be accepted our year, and I couldn't imagine I could make it over a hurdle so high. He insisted I needed to at least apply, and, as was the consistent trend throughout our childhood whenever we disagreed, he ended up being right. I was accepted by CalTech and MIT, but after visiting Olin it became my first choice. During that campus visit, I was part of a group of prospective students that was tasked with building a bridge made of blue foam that we cut with a hot wire. Nothing like that happened in the MIT or CalTech selection process. I sat for interviews, both one-on-one and also in groups. The group interview and the bridge project were designed to give the Olin admissions team opportunities to observe how well we could collaborate. My Olin campus visit was unlike any other, and that was because Olin was unlike any other engineering college — it was a brand-new college. My freshman class would be its first graduating class.

Olin was formed as a learning laboratory where faculty would educate and prepare engineers for the twenty-first century. Undergraduate and postgraduate engineering programs in the United States hadn't changed much since the Cold War, when engineers were expected to be little more than cogs within massive bureaucratic companies serving the military-industrial complex. Olin's founding principle was tailored to our modern era. Its mission was to graduate engineers broadly educated in design, entrepreneurship, and the humanities in addition to engineering.

Olin was committed to training engineers who could bring innovation to market. We needed to be able to identify the needs of everyday people, build the technologies to meet those needs, and then launch companies that would move such solutions into the hands of consumers. And there was this crucially soulful difference: Olin believed integrating the humanities into its engineering curriculum was essential because it wanted its alumni to understand not just whether a solution *could* be built, but whether it *should* be built. We needed to be mindful of ethics and the impact technology would or could have on a society.

Olin initially deferred me — inviting me to attend a year later after doing something out in the world — something that would have certainly benefited me. I still remember the night that path swerved. My parents and I gathered around the answering machine in our kitchen and heard the message informing me I'd been accepted into what would be the inaugural graduating class. I guess you could say that Olin was my first start-up. And as with any start-up, I had no idea what I was in for.

A bit of context for that first Olin class, because it was totally unlike what you would find at any established university. Its unique culture created an oddly wonderful yet at times toxic environment that would forever shape me. First, there was the composition of the class. There were three groups of students. I was one of only thirty-two "true freshmen," who had graduated from high school in 2002. We joined the Olin Partners, twenty-nine students who had already spent a year on campus essentially beta-testing Olin's experimental project-based learning curriculum. Then there were the Virtual Olin Partners. These fourteen VOPs (rhymes with *hops*) had been accepted the previous year, like the Olin Partners, but were given deferred admission. Rather than participate in the year-long beta test of the Olin curriculum, they had spent a year out in the "real world," while occasionally visiting campus or plugging into the Olin curriculum remotely.

While the Olin faculty made a conscious effort to treat all members of our class of seventy-five students the same, and were largely successful at that goal, the reality was that the stratification was baked in. In no time the student body would divide itself, largely self-selecting into those three factions. After all, the Olin Partners had bonded to one another during their intimate and intense year on campus and clung fiercely to their special "Olin Partner" status. The VOPs, by virtue of having spent a year interacting with adults and real-world responsibilities, were significantly more mature and grounded than either the Olin Partners or the true freshmen. Plus, most of the VOPs had developed at least some connection with the Olin Partners because they had been virtually connected during the previous year. What would exacerbate the default into cliques was the palpable, ever-present undercurrent of feelings of uncertainty and pressure.

Every one of us had been accepted to other highly selective universities and had turned down those known opportunities and instead opted to invest our futures in the Olin grand experiment. Therefore, every one of us felt acutely invested in making Olin a successful launch in order to maximize our own successful launches into careers and life. But we had no frame of reference for a "normal" college experience; we didn't have a way to assess whether things at Olin were different, or better, or worse. We didn't know what we didn't know, and as you can imagine, this was a gathering of type-As who most definitely felt uncomfortable without a defined ladder to climb. I think it's fair to say that every one of us was significantly more stressed than if we had attended a more established school.

The first few months were idyllic. Olin's brand-new campus was very much a work in progress, with some buildings still in the final stages of construction. I regarded it as a landscape to be explored. The campus had four core buildings: the hall for the dorms; the academic hall of classrooms and labs;

the student-life center, which would contain the cafeteria; and Olin Hall, which housed our auditorium, along with faculty and administration offices. When we arrived on campus, the cafeteria wasn't quite finished and neither was the academic hall, and so for a while we attended classes and ate our meals in a couple of trailers that jutted like islands from the neighboring college's former soccer field, just as the Partners had done the previous year.

One night during those early days, I snuck into Olin Hall before it was open. Giddily, along with classmates who felt like new friends, we scooted through empty doorways soon to be closed off and poked our heads into rooms we knew we would soon not be free to explore. There was a kind of electricity to being around smart, driven peers all setting out together on this journey. The first night on campus I met my first college boyfriend at a mixer. In the beginning, the student body was a single team tackling the chaos of a brand-new curriculum in a brand-new school. College was living up to my greatest expectations.

While Olin was putting the finishing touches on some of the buildings, the curriculum itself was still evolving. Although our physics and calculus classes tracked close to what I would imagine would have been taught anywhere else, other classes were more experimental. The first programming language we were taught was C rather than, say, Python. (Programmers reading this likely just chuckled to themselves.) Python was designed to be super accessible and easy to learn, while C requires one to master many concepts simultaneously, like memory management (telling the program to remember or forget things), compilation (having the computer process your program before it runs), and detailed and not obvious mandatory formatting (such as ending every line with a semicolon). Teaching a brand-new programmer C as their first language rather than Python is a little like teaching a four-year-old to read with Shakespeare rather than Dr. Seuss. Yet, whether intended or not, even that choice to go with C proved to be a net positive. Our class was still in the united-front phase of solidarity, with giant study sessions sprawled across the lounges; the few students who had reasonable amounts of programming experience took on the role of informal teaching assistants and helped other students, like myself, pick up C. We were collaborating, getting across the finish line together, just like a successful start-up would, and in complete alignment with Olin's master plan.

In those blissful early months, I realized how limited my social life had been in high school. Liam, the boyfriend I had met at the mixer, took me to my first concert, as in my first concert *ever*. My parents never listened to contemporary music. It was mostly NPR and sometimes classical music that served as the soundtrack in our home and car rides. In high school my interests were my schoolwork and debate; contemporary music just wasn't something I was even aware of. Here I was, a seventeen-year-old college freshman who had no point of reference for any contemporary music.

Liam gave me multiple digital folders with music by Ben Folds, and encouraged me to try to listen to as much of it as possible before the concert. He actually had to explain to me that concerts are much more fun if you can sing along. Ben Folds was on a solo tour and played at Brandeis University, maybe a fifteen-minute drive away. I remember the energy of the crowd and that he played so hard he broke a piano string. I remember how magical it felt when the audience picked up where Ben left off when the string snapped and collectively sang the song for the minute or two until the piano tech fixed it and he could enthusiastically jump back in.

A few months into Olin, the honeymoon most definitely was over. Liam and I grew apart and broke up. The little tremors of the systemic stress that the Class of 2006 were all under gathered into quakes that shook the rest of my freshman year.

In a regular college, you have at least four years' worth of students who can help freshmen understand that there's a wide range of ways to be a college student. The first moment I realized something was profoundly wrong was a month or so into school when Olin hosted a prospective students' weekend for high school students and their families. By chance, a handful of women had dyed their hair neon colors the day before for a video-game tournament they were participating in at a nearby college.

There are few colleges in the United States today where anyone on campus would bat an eye at someone walking by with neon-blue hair, including the Olin College of today. But fear and insecurity also breed judgment, as I learned when a friend turned to me as one of the colorful women walked by and said, "Can you believe them? On prospective students' weekend, of all times. Someone should really take them to the Honor Board for lack of passion for the welfare of the college."

I froze. We were standing by the doors to the cafeteria, there to welcome those prospective students, attired in our more than a little dorky official tour guide outfits. My straw yellow hair hid the fact that I was one of those "thoughtless" students. I had never dyed my hair before and had accidentally messed up the process, so the pink coloring I had tried to apply the night before hadn't stuck.

We did not have older students to tell us to relax and not judge each other. Or for that matter, older students who could set really bad examples and make the Overton window of behavior wide enough that most would have the freedom to explore. People were stressed, and they were scared that anything that deviated from the stereotype of the driven, successful college student they had imagined as high schoolers would jeopardize the success of Olin, and with it their success in life.

Not long after prospective students' weekend, a friend came up to me with news they felt I should know. A female student had been complaining that her ex-boyfriend and I represented "everything that was wrong about Olin." Her antagonism to her ex-boyfriend made sense; she was mad they'd broken up. But I didn't understand what I had done to merit her anger. I knew what her criticism meant, though: I was definitely not cool. I was too eager. I literally stuck out at parties, my height rising above most of my female peers. In reality, I was a typically awkward college freshman. I probably would not have been notable at MIT, but she didn't have that context.

I nodded calmly and thanked the friend for caring enough to tell me. I had already been told this dispiriting news by a handful of people. I didn't let it show, but I was crestfallen. The queen bee of Olin had decided that I specifically was her top grievance (for some reason on par with the ex who slighted her), and she was making sure everyone knew. She was one of the primary social organizers in our tiny school community, and I feared her hostility could make me a pariah. The campus was geographically isolated in the western suburbs of Boston; I had no car, and there was no mass transit handy to visit friends on campuses in Boston or Cambridge. I was more or less stuck on Olin Island, and I did not want an anti-cheerleader reinforcing the idea that I was a problem.

I did the math: In a five-thousand-person college you can appeal to only 1 percent of students and still have a community of fifty people who think you're great. At a fifty-thousand-person state school you would have a raucous clan of five hundred to call your own. At a college of seventy-five students, if you appeal to only 1 percent of them, you're alone.

It was the winter of my college experience. I had to learn how to push on, alone.

When the Class of 2007 arrived on campus my sophomore year, Olin's culture suddenly changed and became something akin to what I imagined life was like at established colleges. I think part of it was that almost all of the seventy-five members of the Class of 2007 showed up on campus together for the first time, at the same time. That, and they entered an environment that already existed.

The previous year had polished my roughest edges and the Class of 2007 was my own age. They didn't have the Partner, VOP, or true freshman hang-ups, and if anything, I was a little more mature than they were because I had been on my own in college for a year. I had lived *a lot* of Olin by then. The number of people I considered friends rapidly increased. The Class of 2007 created a new environment where I was recognized as more than just "a problem." For me, it was a reboot.

That sophomore year was the year I took my first class at Wellesley College, the nearby private women's liberal arts school. It was a seminar on the History of the Family, offered by the Education Department and it quickly turned into a group therapy session for all of us. Sitting around discussing how the concept of the family had changed over time, telling stories about how different practices related to what we had experienced in our own families, I would find myself crying along with my classmates.

Over the course of my four years at Olin, I took four classes at Wellesley. I have countless memories of showing up at Wellesley distraught about whatever had transpired between me and the handful of highly opinionated men in my Olin classes, who always seemed to feel wronged by me in some way. The pep talks from my Wellesley professors, most of whom were strong women who had chosen to teach at a women's college for a reason, were invaluable. They would tell me I couldn't let the bullies win. They would tell me that it was likely there would always be men who feel threatened or who just didn't like that I was in "their" space. They would tell me not to let anyone define my life other than myself. If I wanted the life that I deserved, I had to stand up and demand it.

They could give me that guidance because they had lived it. If you are a woman in a male-dominated space, unless those in charge of that space prevent it, there will be men who disproportionately target you. Many years later I encountered a clever study that helped explain this phenomenon of gendered bullying and harassment. *Halo* was one of the most popular video games of my generation. All players (or rather, their avatars) wear identical mechanical space suits as they shoot their way across an alien world. Two researchers, Michael Kasumovic and Jeffrey Kuznekoff at the University of New South Wales, had done a study with three experimental conditions: In one, a prerecorded male-sounding voice said things like "Watch out!" or "Nice shot!" or "Behind you!" In another condition, a female-sounding voice said the same things. And the third condition was a control group where the player only communicated with the same phrases typed into the group chat. To standardize the experience, the same player would play all three of the game conditions. The researchers then recorded what the other players said to the supposedly male and female voices or the control.

When I told my mother the results she laughed. Unsurprisingly to any woman who plays video games, the female voice received the overwhelming majority of the harassment. But that's not what triggered my mom's chortle—it was who was dishing out the abuse. Strong male players (meaning, the most talented at *Halo*) did not harass or bully women. It was the mediocre and struggling male players who committed the verbal abuse. For a mediocre or poorly performing man, every strong woman they could bully off the playing field bumped them one rung up the ladder. My mother's laughter wasn't about video games; she was laughing at the mediocre men who had tried to kneecap her as she came up in the sciences in the 1970s and '80s.

As I write this, we are in a time when it seems that education has become yet another thing society wants to be transactional. In other words: You train me for a job, I get the job. But what I was beginning to realize at Olin and at Wellesley is that students, and ultimately society, gain something special when education is about learning for the sake of learning. Sometimes it's okay to not know what you'll use knowledge for — that's the point of getting an education, the discovery. I think one of the greatest losses in modern engineering curricula is that many of them are so packed with requirements that it discourages opportunities to connect to the humanities and to *humanity*.

When I look back on my years of working with algorithmic products and the rise of machine learning, the most useful class I took in college was not in engineering, and not at Olin. It was a biogeography seminar at Wellesley on speciation and the geographical distribution of living things. Each week we would read papers and dissect them section by section. Much of the focus of the class was on "selection pressures" and how applying a differential pressure to a population (it could be crickets, flowers, or any living thing) might cause it to evolve in a specific way or to split into two populations.

When you're modifying an algorithm, very rarely can the change be considered in black-and-white terms as a "good" or a "bad" change — there are some subpopulations that will benefit and some that will be harmed. Being able to recognize that you need to think about your user population as heterogeneous is not obvious. Being able to think through how your heterogeneous user population might "evolve" as you change an algorithm is even less obvious.

When I was picking my class schedule, did I know that it would benefit me professionally to learn how to think about heterogeneous populations and how differential forces dynamically change them? Of course not. I had no idea of the challenges that I would confront at Google, at Pinterest, and ultimately at Facebook. But isn't that the point of education — to prepare you for problems you haven't encountered yet? It was only because of my unique education at Olin, Wellesley, and even debate camp that I was equipped to see and appreciate how damaging the algorithms at Facebook were. Without that education and the critical humanities-based thinking I developed, I would not have been prepared to spot the toxic influence of the spread of extreme content.

———————

By the time senior year rolled around I had come incredibly far from who I had been sophomore year, let alone the seventeen-year-old I was when I arrived at Olin. All that was left was completing my capstone requirements in order to graduate. For the Class of 2006, we had two final projects. There was the Senior Capstone Program in Engineering (SCOPE) and one for the Arts, Humanities, and Social Sciences (AHS). The AHS was something akin to a minor at most other colleges. When you declared your AHS concentration, it was meant to ensure that the Arts, Humanities, and Social Sciences classes you took would interact with each other in a coherent way. I had decided at some point during my tenure at Olin that I wanted to focus on Cold War Studies.

My AHS capstone was born of the final project I did for my seminar on the History of the Family at Wellesley. In the process of research for that project, I had stumbled upon an amazing article about the early years of the Cold War by Dee Garrison called "Our Skirts Gave Them Courage." The article detailed how back in the 1950s and '60s, in response to the growing nuclear threat from the Soviet Union, the United States had begun to conduct drills throughout the country. They were part of a program called Operation Alert intended to prepare Americans for a cataclysmic nuclear war. The government would simulate a nuclear attack, and citizens would practice the government-recommended plans. These annual "simulations" (and I put this in quotes because the public never really did have to reckon with the true aftermath of a nuclear war) were intended to build public confidence in measures being taken to make the civilian population more resilient to things like nuclear fallout. Back when the program began, nuclear war was likely "survivable" in some general sense. There weren't that many bombs, and the bombs weren't as powerful as they would later become. Taking concrete steps like sheltering underground really would save lives.

Not all Americans were on board with this. For some, any effort to make nuclear Armageddon more palatable was equivalent to inching us closer to oblivion. And as far as these antinuclear people were concerned, Americans shouldn't be encouraged to think they could survive. Instead, if they really wanted to be safe, they needed to demand de-escalation. Groups like the Quakers, Catholic Workers, and Socialists began protesting Operation Alert. The police didn't care that Dorothy Day, leader of the Catholic Workers, was a senior citizen when they beat her with billy clubs. They didn't care that any of these parties were peacefully resisting. You either stood with America and apple pie and prepared for nuclear war, or you were with the enemy.

In that article from an obscure book in the Wellesley library stacks I found the story of a woman named Mary Sharmat. In 1959 she had been struggling with how to respond to that year's Operation Alert. She had watched the increasing militarism of the late 1940s and '50s, and after she gave birth to her baby, she felt she had to take a stand. She believed nuclear air-raid drills taught people to fear and hate the enemy; she was unwilling to hate an unknown enemy when all nuclear war would do is decimate her New York City. She saw only one path forward. Rather than prepare, she would protest. She took $500 in bail money and, given the era, left a roast in the oven to feed her husband, just in case. When the civil defense sirens blared, she went outside and sat on a bench near a civil defense van. The air raid wardens recruited a police officer to force compliance, but Mary insisted she was just fine where she was, thank you, and the cop left her alone. The drill ended anticlimactically; Mary went home to her husband.

That same evening, after the drill had ended, Mary picked up the evening paper and saw a story of another woman who also couldn't silently live with the nuclear theater and propaganda any longer. Right there on the front page of the *New York Times,* Mary saw a large black-and-white photo of a woman named Janice Smith in the park outside City Hall in Manhattan. The photo was so good it looked as if it might have been staged. Just as Mary had done that day, Janice sat quietly on a bench, but instead of being alone, she had her infant in her arms and her toddler standing on the bench beside her. A police officer loomed above her, billy club in hand, in the international posture of "Just try me." Janice looked afraid yet resolute.

Mary had stood up for what she had believed in that afternoon, but nothing she had done made her feel like she had made a dent in the universe. Confronted with Janice's image in the newspaper, Mary no longer felt like she was the crazy one. Seeing even one other person live in the truth made her confident that they were onto something. You never know how your own embracing and living in the truth will impact others — it might be the match that lights a fire in someone else. If they got together and other mothers joined them, maybe then their protest would change something.

Mary picked up the phone directory and started dialing. This was the 1950s, and Janice's number was listed under her husband's name, Smith. There were tens of thousands of Smiths in the New York City phone book. Mary was so determined that she dialed for days before she reached Janice. (She later expressed gratitude that Janice's husband's first name was Jack, because at least it had been in the first half of the alphabet.)

The two women began coordinating, and after finding two more women in the Bronx who had also resisted, they formed a group that successfully led a five-hundred-person protest the following year. The year after that, civil disobedience against Operation Alert spread nationwide, and hundreds of people were arrested on the East Coast alone. The year after that, Operation Alert was canceled. The article I read laid out these events and made a fact-based case that because women like Janice and Mary had weaponized their middle-class motherhood, they had been able to overcome Cold War nuclearism. After all, no previous set of protests had accomplished anything concrete. Police officers were quoted in the article saying things like, "I didn't know what to do. Do you take the baby and then beat the mother?"

For my AHS capstone, I decided I wanted to dig deeper into civil disobedience against civil defense and resistance against nuclearism more generally. Our AHS capstones were only one-semester projects, and I scheduled mine for the fall semester of my senior year so I could visit the Eisenhower Presidential Library for a week over the previous winter break. I drove to Abilene, Kansas, and embedded myself in the archive of Eisenhower Administration papers. It was in Abilene that I was first acquainted with the tools that archivists wield to enable academic historical research. I found government documents like the annual reports on how "effective" each region's preparations were and how many "died" each year in the simulations, based on things like the weather patterns that day and estimated fallout. But what I found most valuable was that I was able to read the real-time cabinet meeting minutes of President Eisenhower and his advisors as they worked through how to deal with the growing threat of nuclear war. It's one thing to read a report; it's another to be a fly on the wall as historical figures discuss it.

This intense library research would be of immense value to me as I collected my disclosures from Facebook. It was because of this extensive research project that I thought about the need for tracking provenance (where documents come from) and contemporaneous context. Part of why I would keep track of the order in which I captured my documents (those numbers next to the folders) was that I wanted people to get a sense of how I was digging through Facebook's file system. I wasn't cherry-picking things. There were clear arcs in what I was searching for and what I was making copies of. I made sure to capture the comments on every document because I saw the value of being able to read people's real-time assessments as they digested information in context. I wanted the world to see that Facebook employees steeped in these documents thought they were accurate. But we'll get to that later.

While I was researching my capstone, those cabinet minutes became particularly important when I later visited the Kennedy Presidential Library in Boston. Operation Alert was instituted under Eisenhower but ended during the Kennedy presidency. If I wanted to understand why Operation Alert ended, that's where I would find the answers. Or so I thought. Surprisingly, there was no mention of resistance to Operation Alert in the cabinet minutes. If the protests that had begun in the Eisenhower years reached critical mass and influenced President Kennedy, one would think there would be some mention of them in various documents dealing with Operation Alert. But there was absolutely nothing about the protests noted at the presidential level.

However, what was obvious in the annual Operation Alert readouts was that in the time between when the program began and ended, the world and the methods of nuclear warfare had changed dramatically. When Operation Alert began in 1954, nuclear weapons were delivered by planes navigating across the North Pole from Russia and were still small enough that if you lived in a suburb and the core of your city was bombed, you were probably going to survive. This is part of why we built out so many suburbs in the 1950s and '60s. Hours would pass between when the planes were first detected and when they arrived—you had time to get ready. By 1962 we had entered the world of hydrogen bombs on missiles launched from submarines. Hydrogen bombs could be three thousand times more powerful than the bomb dropped on Hiroshima. Now you had only minutes to respond to a warning, and your city was going to get flattened.

In my thesis, I presented a graph that charted the estimated "deaths" for each year that Operation Alert took place. The graph didn't have very many points. There were only eight years of that program, but the trend was obvious. The deaths were hockey-sticking. There was no point in simulating nuclear war drills any longer. The protesters had been right: If a nuclear weapon was coming your way, the only thing you had to look forward to was death.

It's easy to look at this outcome and see grounds for nihilism. All of those women put in countless hours and took real risks to themselves and to their children, and you could say it was all for nothing; the program was going to end anyhow. But the women who watched Operation Alert's demise thought that their actions had percolated up to the upper branches of government. Regardless of the reality, what they saw was that when they came together, when they took a stand against role-playing national Armageddon and against the suicidal jingoism that created it, the drills ended.

Their sense of accomplishment—even if illusory—pushed them to do more. The women who led the Operation Alert protests went on to become pivotal leaders in the antinuclear movement and played a critical role in passing the Limited Nuclear Test Ban Treaty in 1963. Nuclear testing may sound innocuous, but a report from the US States National Cancer Institute quantified that the radiation from atmospheric nuclear bomb tests in Nevada from 1951 to 1962 exposed millions of American children to 50 to 160 rads of radiation. The *New York Times* reported that Federal rules for nuclear power plant accidents call for taking protective action when the dose to human thyroids is anticipated to reach 15 rads.

The women who had stood up to the forces behind Operation Alert kept standing up, kept speaking out. There is no denying that their efforts saved lives and prevented generations from being afflicted by illnesses born of radiation. Learning about these women, doing that research, finding my own voice, and choosing to stand up for myself at Olin, as those women had stood up—that was part of my education and became part of who I was to become. Our actions impact other people—we are interconnected. When we choose to live in the truth, we make space for other people to step into the light as well.

———

I walked across the stage at graduation radically more confident in who I was and what I could do than when I had navigated the winding driveway to the Olin campus four years before, even if my future held very little certainty. Most people think of college accreditation as the line that divides degree mills from legitimate educational institutions, but few think about the process by which colleges become accredited.

We were warned before attending Olin that when we graduated our degrees would initially not be accredited. This was always immediately followed with the reassurance that soon after the Class of 2006 entered the world, Olin would complete the accreditation process and our degrees would be grandfathered in. Many of my classmates rightfully anticipated the challenges of conducting a job search during that narrow window with a degree from an unaccredited institution, and chose to apply to graduate school to avoid that uphill fight. Olin was well known in the academic community, even when I was in college, for being a rigorous and selective school. If I had applied to a technical graduate program, I almost certainly would have been admitted. I had high test scores and a 3.8 GPA—high for an engineering major. I had served on the Honor Board and been an officer on the executive board of our student government. I had also turned the college yearbook into a profitable enterprise, which was no small feat in a school with only three hundred students. The problem was that I didn't see myself as an engineer, and I saw getting a PhD as a slippery slope to becoming a professor like my parents.

By chance I had attended an open house at Harvard Business School about their new initiative to increase their number of early career applicants. I had no idea before then that business school used to be like law school in that many if not most students came directly from their undergraduate degree programs. I applied not expecting much—I had four and a half years' less work experience than the average candidate.

The letter that arrived that April effectively threw me onto my fallback plan. I had always expected that if I didn't get in, I'd just return to Iowa and attempt to start a company from my parents' garage. I sat there reading the letter and not really knowing where it put me—I had neither been accepted nor rejected. The letter said I should "come back in two years, we're holding a space for you."

I emailed the admissions team immediately. The letter seemed to assume I had a job lined up. They clearly hadn't considered that a student good enough to make it through their admissions gauntlet hadn't found a way to get a job. Like so many times before, I had unintentionally found a novel way to be an edge case.

"Don't worry," they replied, "Harvard Business School loves entrepreneurship."

I slumped in relief in front of my computer. It's remarkable how fast we reanchor our expectations. In just a few days I had become attached to the idea of getting an MBA. This email was proof I had a future to look forward to. I hadn't messed up my life by attending an experimental college.

For the first time in four years I knew where my path would lead me from here. Or at least that's what I thought as they called my name and I took my seat after receiving my diploma.

Little did I know what was to unfold that summer.

CHAPTER 4

The Googleplex

It's clear there's a lot of room for improvement,
there's no inherent ceiling we're hitting up on.

—Sergey Brin

I stood before the computer help desk inside a small room on the ground floor of the building that would be my workplace for my first year at Google. It was October 2006, my third day on the job, and I faced a middle-aged man who, I feared correctly, was a harbinger of a male-dominated culture that lay before me. For years I had dreamed of working at Google, but after an interview process that can only be described as disconcerting, I had accepted their offer grudgingly and with deep ambivalence. I had no other options. Little did I know this would be the crucible where I would have so many extraordinary experiences and for the first time in my life feel like a whole and autonomous adult.

That year, 2006, less than 13 percent of Google's technical staff were female, and blonde women might have made up 1 or 2 percent, even though there were plenty of blond men. Blonde women were on the sales team or were marketers. They did not code—but I did. On the Google campus for my first full day of work, past orientation and the limbo of waiting to be assigned to a team, I was now discovering what it was like to be part of that statistical anomaly. My Wi-Fi wasn't working, and I had asked the man behind the IT desk for help. I had trouble believing that my fellow engineers would have been met with his body language and sing-song paternalistic tone.

"*Wooould* you say that you're more of a wired or unwired person?" he said leaning in, head slightly tilted to the side, very clearly trying to be approachable and helpful. He smiled broadly at me and waited.

Technologically speaking, this was still the era when people would sometimes opt for the inconvenience of plugging their computers into a wired Ethernet connection to get internet. It was faster and more consistent. I paused for a moment before I responded. Some of my fondest childhood memories flashed through my mind: Helping my father rewire our home (twice) for data, once for Apple's proprietary and ancient AppleTalk standard, and later on for Ethernet. We drilled into the interior and exterior walls of our home and snaked wire-filled conduit outside the house and between the studs of internal walls. It felt powerful to hack our home. I am also an electrical engineer by training. At Olin I built radios from scratch and studied encoding protocols for how to translate information into the signals that could be transmitted through the air or down a wire.

"I'm going to be moving around between meetings," I responded matter-of-factly to Mr. Help Desk. "I need my Wi-Fi to work." I decided the less context the better in order to get him to take me seriously.

When you talk to people about a job they really want, they often say things like, "I just *have* to get that job." When I uttered that line about getting a coveted slot with Google's associate product manager program, which was then the crown jewel management rotation program at the company and perhaps the entire tech industry, I meant it with an intensity that may be difficult to comprehend. This was not a job. This was my ability to escape Iowa, for real this time. This was my chance to have a future. And by this point in my job search, after realizing that The Market did not value an engineering degree from a new unaccredited college, it felt like one of the only opportunities I was going to get. It was my life preserver, and I was going to seize it and cling to it with every last drop of energy I had.

A few months earlier, around July, I had reached out to the admissions department at Harvard Business School to discuss the process for getting my start-up plan acknowledged as a formal work experience. My deferral letter had given an October 1 start date as the latest I could start working, and I didn't want any surprises when the date drew near.

There was ominous silence for a few days, and then a response appeared in my inbox that made my blood run cold. "After greater consideration, we would like you to have a more structured work experience." I stared at that word, "structured" — they wanted me to get a job job, *and* be working, by October 1. Did they understand how hard that was this late in the game? For those not familiar with post-collegiate hiring, freshly minted engineers are like Play-Doh. They're relatively interchangeable and need a substantial amount of shaping. Many companies will not hire an engineer straight from college because the initial investment is so high, which relegates the work of onboarding brand-new engineers disproportionately to the largest companies. Those same companies often have the most formal and rigid hiring systems because they have to process such large volumes of applicants. As I began to apply for job after job, I became familiar with the dreaded phrases "Must have accredited engineering degree" or "Apply to start June, 2007 or later." A year from now in other words. Long after my October 1 HBS deadline.

I was saved at the eleventh hour, in late August, by a delayed response from a Google recruiter. Google was still young enough that it could be flexible. Some of their best engineers were high school dropouts. They had sufficient profit margins that they would rather sift through a mountain of résumés than miss jewels in the rough. I had an internal-to-Google recommendation from a friend of one of my Olin professors, and that was enough for Google to at least have someone call me for a phone screen.

I had applied at that point to at least fifty-five jobs and gotten three phone interviews — one of which was secured by the HBS admissions team. I *had* to get this job. It was one thing to say I wanted to do a start-up when I had the illusion I could just get a job if it didn't succeed. After having tried to get a job and not gotten very far, I felt I couldn't risk losing my HBS deferral. It was Google or bust. There was no Plan B. This probably wasn't an irrational fear — one friend from my class with high grades was still without a job two years after graduating.

A week later I had made it through the phone screen and was told to be on a plane to Mountain View, California, two weeks later. I knew I had to stick the landing. A standard level of preparation for interviews might be to spend a couple of hours reading up on what the press is saying about the company, but I didn't want to leave anything to chance. I knew I could not control what happened in the interview room, but I could control every moment before. Over the next two weeks, as the clock ticked down to my moment of judgment, I invested at least a hundred hours preparing a portfolio of ten product ideas that Google could build. I included an explanation of the current unmet user need and what might be built to meet that need, along with a rationale of the business value Google would derive. I wanted to demonstrate the value of my multidimensional Olin education. I wanted to show, not tell, that I wasn't afraid of working hard. I was hungry, and I wanted Google to know it.

During my three years running the Olin College yearbook, I became skilled at crafting polished documents. For my onsite interviews, each interviewer received a glossy folder with my résumé and those ten product ideas printed on glossy cardstock, bound together like a booklet with three-hole punches and brads. My vision was clear: This was my dream job. I was going to get my dream job. When I flew out to California for the in-person interviews, I stayed with my aunt and uncle in Palo Alto. One of the many ways my aunt showed her support was driving me down to Mountain View and dropping me off that morning in the parking lot of the Google campus.

For the majority of onsite interviews at that time, Google would bring in a candidate for the day, do three interviews, and decide if the candidate was worth bringing back to campus for another day for three or four more interviews. This was easy because few candidates were flying in — the pool of available talent was large enough in the Bay Area to meet Google's needs. Evidently I had done so well in my first three interviews that day that Google recruiting decided to get right to the second round. I was doing an end-run around their standard playbook by coming in from so far away, and my recruiter was left playing catch-up as he tried to piece together a slate of interviewers of escalating seniority to produce a packet of feedback sufficient to submit to the infamous Google Hiring Committee.

Back then there was so much interest in working at Google that one needed a perfect slate of interviews to get hired, or six iridescently glowing interviews to offset a single negative or neutral review. Later, after years of research, Google would find that surviving this gauntlet of interviews did not improve the quality of the hires (as it turns out, you gather sufficient signal with the first four interviews; you don't need six), and that every incremental interview provided an opportunity for the interviewer to "not click with" an interviewee, and thus lead to them getting "dinged." What was originally conceived as a quality filter ends as a diversity filter—if you aren't substantially like the people interviewing you, you are more likely to be eliminated.

My friendliest interview that day was my engineering interview. The engineer asked me how I would build Image Search if I started from scratch. It was the only algorithms question I was asked, and probably one of the few I could have answered, given my analog electrical engineering degree. By totally random chance, during my summer studying abroad at Uppsala University in Sweden between sophomore and junior year, one of the two classes I had taken focused on data mining and information retrieval. This set me up perfectly for when the engineer asked, "Imagine you were building Google Images from scratch. Would you rather do breadth first or depth first search?"

This is a great interview question. It highlights how different choices about data and algorithms can yield very different experiences for the user, and I think it marks the first time I ever thought about the intersection of algorithm design and user experience. "Depth first" means you try to get every image from a single website before you move on to the next one. "Breadth first" means you get one or a few images from a website and then move on to the next, with the intention of coming back to all those previous websites and getting incrementally more and more images in subsequent rounds. "I would do breadth first," I responded. "In a search engine, you want a wide variety of possible images to show a user, and the images within a website are less diverse than the images between websites."

I remember the pride I felt when the engineer responded positively to me as we moved from step to step in the scenario. It was the only interview I felt confident about when I went home that evening. In all the others, no matter what I said, I got the impression I had missed the mark—interviewer after interviewer with a flat affect and a mild scowl, occasionally looking over the tops of their computers as they took copious notes. Each exuded an almost palpable sense of doubt. Oh, and they were almost all men. In my seven interviews that day, I had had nine interviewers (two shadows had ridden along, learning how to interview), and of those nine, only one was a woman.

In one interview I was asked how I would design a gas pump. At one point I commented for context that in Massachusetts at that time, you had to continuously hold on to the handle of the gas hose—there was no little switch to flip so you could keep your hands in your pockets in the bitter New England winter (because of laws about "safety," of course). My interviewer's head shot up above his laptop screen and he glowered at me as if I were making things up. *"Really, now?"* he seemed to be asking. His shock was understandable. When Massachusetts finally relented to common sense in 2015, only one other state, New York, still banned "hold-open clips" on gas pumps.

One vice-president picked up my lovingly created portfolio when he arrived, and unlike the other interviewers, who had not even acknowledged it before asking me questions, he took a moment to quietly flip through each page before talking to me. With a disapproving stare, he said, "These are all just search verticals," before dropping it back on the table. I didn't know yet what a vertical meant in this context, but his dismissal deflated me. I didn't know it at the time, but he led one of the few divisions of Google that wasn't focused on Search.

Because my slate of interviews was so haphazard, I was left alone in a small interview room, maybe seven or eight feet square, sometimes for as long as an hour, unable to relax because I didn't know when the next interview would begin. The high industrial ceilings made it feel even narrower and more claustrophobic. I was not allowed to leave the room unescorted, not even go to the bathroom. This was before smartphones; I was just left hanging alone with my thoughts, awaiting my fate.

More than once I had an interviewer show up and ask me how I was doing. I would offer versions of, "Good, just tired. I've been here for, like, five or six hours already." Instead of responding with anything resembling a human-to-human response, like "That's a long day," they looked at the paper they'd been handed, then looked back at me, head cocked to one side and said skeptically, "I'm the only name I see on your interview list." Like I was trying to pull a fast one on them by saying I was further along in the process. It was clear that Google didn't often cram two days of interviews into one.

I left Google that day not wanting to return. I had walked onto the campus feeling like a starlet arriving on a sound stage. I was in paradise. I had made it to the mother ship. It had taken only that one day of interacting with Googlers in person (or rather, as I would later learn, their interview process) to sap all that enthusiasm right out of me. What I really wanted after my time on the Google campus was an offer from the Boston Consulting Group (BCG), which was the only other on-site interview in my pipeline. I had already interviewed with their offices in Chicago, and now I had scheduled a second round of interviews in Boston.

Two weeks later, on my birthday and the night before my Boston interview, I got a phone call around 9:30 from Google, where it was only 6:30. They wanted to hire me, and I should check my email for the offer letter. I couldn't get to my computer fast enough.

The next morning, the excitement-fueled insomnia that trailed from Google's email took its toll. I could feel the exhaustion holding me back from performing at my brightest in front of a string of consultants, each of whom I wished I could be like when I grew up. They were confident, warm, and clearly good at what they did. When BCG declined to offer me a job, I accepted Google's offer with a feeling of resignation. At least my period of uncertainty was over. Worst-case scenario, my Harvard Business School deferral was now safe. But what was I getting myself into?

―――――――

About five weeks later, my father and I hopped in my station wagon and drove out to California so I could have a car in the expanse of suburbs surrounding Google that I thought of as burblandia. I moved in with my aunt and uncle for a couple of weeks while I figured out where I would live. I was back on the Google campus, only now as a full-blown new Googler, or in Google parlance, a Noogler. Although I had no way of knowing it at the time, my Google full-time offer had come at exactly the right moment to set me on my course to becoming an algorithmic specialist, and later a Facebook whistleblower.

Looking back today, I hope the story, or rather stories, of my initial experiences at Google may help those who feel lost in their early careers. It may help them envision how even rocky beginnings can lead into deeply fulfilling life paths if they're willing to persevere in hard times. You can feel (or even actually be) profoundly behind and still catch up with persistence, and if you can be kind enough to yourself about the process to persist.

My first two or three days at Google were spent in orientation, being walked through health and retirement plan options and being generally briefed on the wondrous benefits Google provided that would allow me to meet all my needs on campus. Need an oil change? You're covered, there's one in the parking lot. Hurt your knee playing basketball? No worries, there's a doctor and world-class massage therapists. Hungry? Three gourmet meals a day, and a smoothie bar. Dirty clothes? Free washing machines, we'll even throw in the soap. Feel free to work as much as you want — we've got you.

At the end of the second or third day, as orientation was wrapping, I was picked up by Diana, the coordinator of the associate product manager program who reported to Marissa Mayer, VP of Search. I remember the delight I felt at seeing Diana; she had been my lunch interview when I had come out to campus earlier. Sitting in Pacific Cafe with her, eating unlimited free sushi and chatting, had been one of the few happy memories of my day of interviews.

My first rotation was on the Google Ads team, and my first desk was in a hallway on the first floor of building 42. I joined Google right on the cusp of their first expansions away from the core Googleplex buildings. There was still just barely enough room to keep the Ads and Search teams housed in the original central buildings known as 40, 41, 42, and 43. But "just barely enough" means sometimes you have to wait for people to go on vacation or change jobs in order to reach even the lowest rung on the ladder. In other words, getting to sit in an actual cubicle. Until then, I would sit in the hallway.

This was my trial-by-fire introduction to the delight of an open-floorplan office. At my hall desk I always had the feeling I was being watched, as it was located in one of the main halls between the Ads product manager cubes and the building exit. As someone who craved validation, knowing that anyone could walk up behind me and judge my activity or inactivity was an agonizing introduction to the working world. In the month or two I sat there, I never tuned it out.

Google was a giant wake-up call because until then I had never really felt what it was like to be routinely sexualized—I had never felt like people were looking at me as an object. My only other frame of reference was Olin, and its size and composition made it a world away. Admissions were equally divided among men and women. I was just one of half of the student population, and it was as if we were a small tribe and everyone was a sibling. Perhaps more importantly, there were two thousand other undergraduate women five minutes away at Wellesley. Now I was among the 13 percent female minority within the technical staff at Google. I distinctly recall during my first week a male colleague standing next to me in the cafeteria food line sizing me up with zero subtlety. He scanned me from head to foot and back up again and nodded in approval, like he was considering me among the lunch options.

I've thought back on why it hasn't felt that disruptive to move into the public eye, and I've come to the conclusion that I never felt I had privacy when I worked at Google. People knew my name who had no reason to know it. I was a low-level junior employee, yet when I would sit at an outside table at any restaurant on Castro Street, the main drag of Mountain View and the only place with a density of restaurants, I would hear a stream of my name repeated in greeting, sometimes from people I recognized and often from people I didn't.

One of the most interesting analyses I've read on making the working environment less toxic for women in tech relies on a very basic mathematical argument. Let's say 1 percent of men (or less!) are the problem when it comes to sexual harassment or active misogyny in the workplace. When you go from 13 percent of your workforce being women to 25 percent, you halve the torment any given woman will receive, because that 1 percent of assholes now spreads their attention over twice as many women. If you raise the inclusion rate to 35 percent, you reduce harassment by 63 percent. This is another rare situation where the Reagan Administration's EPA ethos applies: The solution to pollution is dilution.

In 2006 there were vanishingly few women leaders at Google, but somehow by the stars' alignment, I was assigned a woman manager. At first, as I was transitioning into this overwhelmingly male environment, I had a very hard time adapting to being stared at or objectified. I remember breaking down in the office of my manager, Amy Chang, a month or so into working at Google because the constant strain of being sexualized was that foreign and stressful. She handled it with such composure as she sat and listened. Sometimes there is nothing that can be done about an unfair situation other than being a good listener. Compassion goes a long way in helping others to endure.

My life had changed so rapidly and profoundly that I was willing to go to great lengths to ignore all the parts of Google that were difficult. Three weeks after I started came the day when I reached the milestone I had been dreaming about since I was thirteen years old: I got my first paycheck. I was now financially independent; I was now free. I sat at my desk, still in the hallway, and went shopping online. I had developed a love of second-hand shopping in college, stretching dollars as far they could go. I still over-index on "previously loved" clothing, but I remember the thrill and feeling of power of pulling up my favorite shopping website and buying new clothes for work. I spent maybe $400 in 2006 dollars, but it felt phenomenal. It felt like liberation.

I had also merged into the social scene of the associate product managers (APMs) in a way that I never had before. Because Olin College was brand-new and none of its graduates had yet moved to the San Francisco Bay Area before I did, there was no local alumni network. I appreciate now how important it is for graduates fresh from college to have people to reach out to when they emerge into adulthood in a new city. I had no one, and as a result merged directly into the community of APMs and by proxy into a network of Stanford alumni. As someone who had always been on the periphery of the social scene in college, and at a school that was physically isolated in the Boston suburbs, it felt like paradise to go to at least one house party a night every Friday and Saturday. Often these parties were hosted in palatial multimillion-dollar houses nestled into the peninsula suburbs, places that were clearly owned by the host's parents. Exposure to the bounty of Silicon Valley made it easy to turn a blind eye to the dysfunctional parts of the valley's culture and start-up ecosystem. If you played the game, this could be yours.

Google was at its peak period of excess. It had IPOed just two years before I joined, and it had acquired YouTube a week before my start date. During my first weeks at Google, a construction crew removed a giant glass wall in the building next to mine in order to bring in a one-to-one scale model of SpaceShipOne, the first privately funded crewed spacecraft to reach outer space and the winner of the first XPRIZE. It was a piece of art the company bought or perhaps commissioned to inspire us to believe there were no limits to what we could achieve. It hung prominently over the main entry staircase in the headquarters building where Larry Page and Sergey Brin kept their offices. I would walk under SpaceShipOne almost every day on my way to my favorite Google cafe for an endlessly changing feast of gourmet dishes. Fascinating and famous people were always showing up on campus to give Tech Talks. It felt like the Googleplex was the center of the world, and that we were special for having braved the interview gauntlet to be awarded our places there.

I wanted to be in that domain of otherworldly success so badly, even if getting on my feet at work was proving to be a bumpy process. I was a junior product manager—a bit of a software jack-of-all-trades. Software product management requires identifying problems users face, building consensus with your team to identify and prioritize solutions for addressing those problems, and then helping the team develop and execute a plan for building those solutions. One role of a product manager is to fill any gap that emerges, or to find someone to fill it. Another role is to absorb uncertainty so the team can move forward faster. You don't need to wait for that last holdout to agree, as long as the product manager is willing to accept responsibility for what happens if things go wrong.

I didn't know it at the time, but almost all associate product managers face a rough patch at the start of their careers, to the point where a surprisingly large number of them wash out almost immediately. For the APMs straight from college, most of them were at the top of their class at elite

Computer Science programs. They were young adults who had been near-perfect in high school, and then again in college. Now we faced a profoundly new challenge — a position in which almost none of us would be able to do a good job for the first months or even the first year we were in it. The process of product management is nuanced, and has many subskills. Almost all APMs hit a point maybe six months into their first role when they have learned enough to realize how profoundly bad they are at their job, and most challengingly, they also realize there is nothing they can do to get better, other than simply persist. The only way out is *through* — you're only going to get better by continuing and continuing to try.

My class of APMs was the last in which the vast majority of the year's cohort did not enter straight out of college — most of my peers had previous work experience. And the difference in experience levels was meaningful. About a third of the twenty or so people in my class were twenty-seven or twenty-eight years old, and entering the field from other careers. Another third was twenty-four to twenty-six, whereas I had turned twenty-two less than a month before I joined.

I know. I know. I can hear the tiny violin playing. But for me my mini-career crisis was excruciating because I had never before experienced being significantly worse than my peers in any activity I had ever cared about. Yes, I struggled to run and as a result I was bad at sports, but my parents never watched them and I hadn't grown up caring about them. This was different. I desperately wanted to be at Google, and be good enough to stay. I was still measuring my own value against others', and in this case it was a comparison that was inherently unfair to me. Starting the year after I joined Google, all APMs were required to join straight from school to level this playing field.

A big part of what kept me motivated — even when I felt shame every day for how poorly I felt I was doing — was the choice I made to control what I could control and to let go of what I could not control. I was not as good as my peers were. But I could choose when to quit. Even if I was slower than them, I could choose to outlast them in the long run. My faith in process and just sticking with it proved to be the right call. Over that first year I got my feet under me, and besides shipping many different Google features and doing a good enough job to close out that year consistently earning "exceeds expectations," I also had a series of experiences through the APM program that allowed me to grow into a confident young adult.

Some of those experiences happened naturally, like the fact that associate product managers were frequently assigned as the notetakers in senior meetings, including the Ads team edition of the product strategy meetings that occurred once or twice a week. I would sit in those meetings and marvel at the topics the most senior leaders of the Ads org hashed out. We sometimes forget that any decision you see coming out of a corporation was made by people, and I had a front-row seat to watch the process of making those decisions.

Note-taking as mid- and senior-level product managers proposed new products or walked executives through the choices needed to resolve their projects was a unique education in the sausage-making of software. We now live in a world where many people are informed about and care about privacy and data usage in tech, and as a result there are now entire compliance departments in most large tech companies for deciding and accurately communicating how your data will be collected and used. But back in 2006, during my first year at Google, there was no such thing. Privacy was a sufficiently peripheral issue at the time that the privacy product manager for Google as a whole was a fellow fresh college-grad APM with less than a full year of work experience. When, after one Ads Product Strategy meeting, I was tasked with writing a draft privacy policy for Google Analytics or maybe even Google Ads as a whole, I met with him, and he walked me through how privacy was thought about at the company and suggested some frameworks I might want to use for my draft. We collaborated on a document, and a few weeks later I brought it back to the strategy meeting for review. I brought a number of copies along with me to help with the discussion. Which, I learned, was something you do not do.

I was thrilled at the chance to present my project to leaders whom I respected so immensely. I proudly passed around the printouts I had brought and sat nervously reviewing the slides I was going to walk through as soon as the assembled leaders finished reading my carefully chosen words. Imagine my shock when a look of horror froze the face of one of the most senior leaders in the room once they realized what they were holding in their hands. "Let's collect these copies," they said briskly before giving my manager and me a lecture in front of the similarly baffled audience. The privacy policy project had not been my idea—a few weeks earlier my manager had brought something in for review, and questions around the privacy expectations of Google Analytics and Google Ads had emerged from it. I had tagged along to watch my manager present and in the process was deputized to bring something back for everyone to discuss.

This was my first exposure to the idea that documentation was *dangerous,* at least to those who fear casting light on their actions. I don't think the leader had any malicious intent. There wasn't some devious thing Google was doing or that they were plotting to do. But the leader recognized that writing a draft policy, even if it was not adopted, could pose a liability down the road if a lawsuit was later brought or circumstances out in the world changed. A draft policy, especially out of context, might be used by someone to argue that Google knew about a situation and had purposely chosen not to deal with it—as in, it was a documented choice not to act.

My manager was in charge of Google Analytics, which when I joined Google was weeks away from a radical transformation. Not so long ago, it was next to impossible to gain even basic insights from the data you collected on your website, but a small start-up called Urchin, led by Paul Muret, redefined the expectations of the industry. Paul and I crossed paths for less than a year during my first rotation at Google, yet he shaped how I think about product design to this day. He recognized there is a difference between most software experiences people interact with on a daily basis (for example, Facebook's News Feed or Netflix), which are designed to meet users' needs with the least amount of friction, and analytical products that are intended to help someone learn or discover something novel from data. One is meant to be consumed, while the other is meant to be, as Steve Jobs put it, a "bicycle for the mind"—to make possible thought processes that would otherwise be tedious or impossible.

Before Urchin, companies had made web analytics software similar to what Google Analytics was when I joined—page after page of charts, as if the volume of charts were proof that the product provided value. The only problem was that the vast majority of people do not understand how to use data. The puzzle pieces they need for a brilliant insight might be buried in those charts, but most people could not knit them together in a way that could drive value.

Paul had flipped the script. He used a technique he called the curricular model of product design to select which charts to show the user, and in which order, so the path of least resistance flowed naturally from your goals to insights you could act on. He literally sat down at his kitchen table one night and wrote what was effectively a college curriculum on how to analyze web traffic, anchored on three main goals: Do you want more users? More page views? Or more purchases? Instead of expecting users to know exactly which chart to retrieve to move from step to step in their analysis, users would start at the top with their goals, and information would only be presented that would help them deduce the subsequent questions that arose from each goal. He wanted to make it easy to draw conclusions that would unlock each iterative step your web business needed to succeed.

This may sound simple now, but it was revolutionary then. Two Googlers, Wesley Chan (PM) and David Friedberg (Corp. Development), spotted the Urchin team at the Search Engine Strategies conference, and before long Google had snapped up Urchin for low tens of millions of dollars in 2005 to help Google catch up in the website analytics space. This acquisition would later be considered one of the most profitable software acquisitions in history. Google knew from working with some of the most advanced online retailers in the world that Google AdWords had the best return on investment a marketer could hope for. Back in the mid-2000s, you could put one dollar into Google AdWords and you would get back many multiples' worth. Unfortunately, all but the largest retailers lacked much data sophistication back then — even those who spent large sums of money on AdWords often weren't aware of the relative value they received from Google advertising. Google Analytics, now updated with an Urchin-inspired user interface and one of the highest performance backends for crunching data available, helped make this value obvious.

Amy Chang, my manager and the lead product manager for Google Analytics, had the insight that if she wanted her team to get the recognition or resources they deserved, she needed to demonstrate that advertisers who used Google Analytics ended up increasing their ad spend. For the tens of millions of dollars in investment, Google received tens of billions of dollars in extra advertising revenue over the coming years. Once the company was able to clearly help people think through how Google advertising interacted with their businesses, advertisers turned around and poured more money back into Google AdWords and AdSense.

The only reason we knew this—and it may sound far-fetched today in our data-soaked and data-driven world—is because Amy fought for a single economist to do this analysis. In 2006, the era of data science as a business practice was just barely forming. The first people in the industry with the title "data scientist" were still two years away—the title itself was chosen through live experiments run at LinkedIn. I remember marveling at the power of data to demonstrate value. Information really was power.

By far the most nerve-racking and rewarding project of my first year at Google began with an email instructing me to show up at a conference room in the building next door, where the core Search teams worked. It seems amazing today, but when I joined Google, Google.com only showed results from websites—a simple list of ten blue links connecting you to webpages that might contain relevant information. If you wanted images, you went to http://images.google.com. If you wanted news, you went to http://news.google.com. Each was a separate search vertical (the very thing the executive had tut-tutted at my interview). If you didn't know what type of information you wanted, you might have to go to several Google-owned sites looking for information. There weren't even consistent links between products—you would have to know that Google Video existed in the first place to be able to check that there was relevant information from videos available for your query. It wasn't just the single search and click away that we expect today.

Google was about to launch a quantum leap in search: Every time you used Google.com, you wouldn't just get back websites, you'd also get relevant images, videos, books, and news articles—each type of content wrapped up in its own tidy OneBox. For the first time, search could be a one-and-done experience. Since the Search Quality APMs were all busy with this launch, Marissa Mayer had reached out to the Ads team for someone to help her prepare for her presentation as part of the international launch for this new product. She needed someone to come up with demonstration queries (words used in the search box to retrieve information) that would trigger these integrated results, and also to help craft her slides in collaboration with an external communications company.

This was my first up-close exposure to Marissa. I scurried away from the meeting to meet with all the relevant product managers working on the Universal Search launch and asked them what they thought would make good demos. Today, if I asked any of those PMs, "Which queries trigger your experience?" any of them would send me a spreadsheet with thousands of queries that could trigger their OneBox in the trials that preceded the launch, along with data that would allow me to quickly mine their mountain of queries looking for compelling narrative gold. The world has changed enough that using data would be an intrinsic part of their job. Instead, each PM I visited rattled off a single digit's worth of search terms that returned a book, a video, an image, or a news OneBox nestled in among the search results.

I was left throwing queries at the system and making my own lists through trial and error. Just like the other APMs, I did not yet know how to retrieve information from the Google search logs. Over the next few weeks the external comms team and I stitched together animated slides along with live Google.com demos into a coherent narrative. Yet that narrative carried certain obvious risks. The nature of the OneBoxes was that they were dynamic—they showed up only if a query appeared to have an outsized image, news, or video "Intent." For example, when someone typed a query like [Charleston], it might yield a map OneBox, because most of the time when people searched for [Charleston] they would directly navigate to Google Maps or other map websites from the search results page. But to my chagrin, search intent could wax and wane. The same query for [Charleston] might yield a News story after a celebrity had visited Charleston or a tragedy occurred there, and suddenly people wanted to know about what was going on in Charleston *right now* and navigate from the search results to Google News or another news site.

This was good for searchers—the goal was to always give the most relevant results—but incredibly bad for me. Marissa was adamant that she did not want to be surprised the morning of the launch or, even worse, when she was on stage. Her criteria were clear: she wanted queries that would not change. For most of the verticals I felt confident we could find stable queries. If you search for [red baron autobiography], *The Red Fighter Pilot* will continue to be the book returned—most people don't write two autobiographies in their lifetimes. But for the News OneBox, it was impossible to predict what would be "newsy" on any given day in the future.

Launch day was rapidly approaching, and with it came rounds of rehearsals and revisions. Marissa announced that as a reward for my hard work, I was going to stand on stage and advance her slides as she moved through her presentation. This would be my first exposure to the international press, and it was both terrifying and exciting. It was not an honor without cost. The level of strain was high. The night before the reveal my phone rang at 11 with a call from Marissa's assistant. I had been balancing my normal workload with the additional assignment of helping Marissa prep, and this meant that I was working very long hours at Google. I had only just made it home after working out when I had to process Marissa's assistant's voice spilling out of my Google-issued BlackBerry as it fulfilled its purpose in tethering me to work at all hours. "Marissa would like you to come in and run through her slides with her."

I was exhausted. As I listened to her I thought, *What would Marissa have done if I had lived in San Francisco, like many of the other APMs?* At the check-in that afternoon, there had been no warning or indication we might need to do an ad hoc run-through in the middle of the night. What about the need to be well rested the next morning? Fortunately, I lived only ten minutes from campus.

I look back on my response as a symptom of the lack of boundaries I had with Google and Marissa at this time in my life. I leveled with the assistant. "I just got home from the gym, do I have time to shower before coming in?" There was silence on the other end of the line.

"How long will that take?" she asked, clearly relaying a question directly from Marissa.

"I can be there in thirty minutes," I said, almost holding my breath. I didn't feel I could say no. Even asking for a few minutes to wash off the dried sweat and deal with my matted hair felt dangerously close to a no. Again, a long pause.

"That's fine, just get here."

I rolled into building 43 around 11:30 to find Marissa, her assistant, and a communications coach who apparently also made emergency middle-of-the-night visits hunkered down in a conference room near Marissa's office. I've since fantasized about what it would be like to sit down with that communications coach for a drink and listen to him spin a yarn or ten about what other Silicon Valley excesses he had encountered over his many decades polishing the public personas of tech executives.

We spent two hours that night running through Marissa's presentation over and over again. Each time, the speech coach videotaped it, played it back for her, and gave her small pointers. One of the most frequent comments I heard after I came out as the Facebook whistleblower and gave my Senate testimony was that "No whistleblower could be *that* good." Usually this was followed by some conspiracy theory about me being a crisis actor or a CIA plant. The truth is, I have been lucky to have a number of unique educational experiences along the way, this among them, often by taking on most unexpected tasks that technically were outside of my job specifics.

By the morning of the launch, I was exhausted but ready to go. We needed to be at Google early so Marissa could have her hair and makeup done. I sat next to her as the makeup artist asked questions about her preferences. As someone who had had so few experiences with traditionally feminine culture, I was transfixed as they discussed the merits of different colors and techniques. Marissa and I have similar coloring, and I still gravitate to a shade of blush I learned about in that conversation. I have always been grateful that Marissa broke the stigma for me that you must choose between being smart and being yourself, even if what you appear like doesn't fit the societal stereotype for intelligence.

We ran through her slides one last time, and then it was time to go on stage. Google had, and pretty much still has, the ability to summon to the Googleplex just about every major press outlet in the world for a big release. Marissa strode onto the stage and I walked a few steps past her to stand behind the podium off to the side. My silver MacBook was already there and plugged in by the techs who had prepared the room. I smiled at the audience and typed in my password while I waited for Marissa to begin. She paced the stage as she walked the journalists through "the need": Google's mission to organize the world's information and make it universally accessible and useful. Google could not fulfill this mission if accessing information required knowing about the existence of each of Google's sprawling constellation of products and predicting beforehand which one of them would contain your information. Universal Search was going to change the way the world accessed information. She walked through the demo, query by query, images, videos, news stories, and books dynamically wandering in and out of the ten blue links that had always been Google search up to that point. You could see that the audience understood this was Big. I felt a surge of pride that day as I clicked through the demos we had crafted. Few women executives in Silicon Valley were as senior as Marissa, and I had done my part to help the world see this new realm of search she had led the team to summon for the world to use.

In the year that followed, I had several similar opportunities to work with Marissa for talks she gave at conferences or annual events, like the announcements of the Google Search Queries of the year. It would be years until I would work personally with world-class speech coaches during my prep in the days before my *60 Minutes* interview, but during those early years of my career at Google I logged many hours watching a world-class communicator work with some of the best teachers in the business.

The summer after my first year at the company brought an incredible opportunity: the associate product manager trip. It was an around-the-world junket lasting two and a half weeks and zigzagging from Tokyo to Beijing to Bangkok to Bangalore to Tel Aviv and back home. The APM trip was a tradition. What had once begun as a five-day jaunt to Europe now sprawled across multiple continents and required a full slate of vaccinations for tropical diseases — the kind that left a number of APMs bedbound with fevers after the Google doctor had jabbed us in the arm one by one.

We took off from San Francisco's airport through the international terminal, and the dramatic soaring arch of the roof mirrored how I felt setting off on my first adventure circumnavigating the world. We were joined on our trip by Steven Levy, a veteran tech journalist who had been covering the movers and shakers of the industry for decades. I did not know it yet, but Steven would offer me some ballast on the trip, as he was not a twenty-something man jostling and posturing with his twentysomething male peers in our own mini frat house as we roamed the world. Nor, certainly, was I.

We landed first in Tokyo and spent three fourteen-hour days being exposed to experiences as diverse as eating a breakfast of fresh sushi at the famous Tokyo fish market only feet away from the thousand-pound frozen bluefin tuna we had seen auctioned minutes before, to rooting through little stores of quirky electronics in Akihabara in a mall that consisted of story after story of similar hole-in-the-wall vendors. We walked the gardens around the Muji Shinto Shrine under transparent umbrellas as the remnants of a cyclone washed over the islands, dropping a heavy rain that followed us for the rest of the day. A moderate earthquake shook us, and the next day we heard warnings of a nuclear reactor that was leaking slightly as a result. I was so far from Iowa. So far from home. Working for Google. I loved every second of it.

In 2007 the mobile internet was just becoming a thing in the United States. I remember my first week at Google and the way the other APMs had clued me in on how to get my hands on a corporate BlackBerry with a comped data plan—a pricey luxury I had never indulged in before. I remember how magical it felt to have Google Maps available to me as I drove, even though it would barely be recognizable today. Years before the era of phone turn-by-turn directions or even GPS, it felt like magic. Before we had the mobile internet, you'd print your directions from MapQuest or Google Maps before you left home and carefully hit the trip-distance button after each turn so you could carefully calibrate the tenths of a mile before your next turn. Pay close enough attention, and you wouldn't overshoot it. Mobile data meant that if you missed your turn you hadn't driven off the edge of the map, and the confidence that came with that knowledge felt like a superpower. I would roam free and wild across the Bay Area, thanks to a fake "GPS" blue dot that showed me approximately where I was. Living in a world where wristwatches now have meters-accurate GPS, it feels like fiction to write that I felt set free by a cell-phone app that crudely triangulated my position from the cell-phone towers around me.

Japan was completely different. That nation had hopped on the mobile data train years before, and that was part of why we were there. The cities included on the annual APM trip changed from year to year. Part of the perk of serving as a host for this trip was that at a minimum Marissa was going to show up in your office, and your site lead was guaranteed to get some time with her. But Tokyo was always on the itinerary, because this was the only place in the world where you could see the future of the mobile web. Japan was meant to inspire us about what technology would come next to the United States. Each of the other cities on the trip, on the other hand, was scheduled to remind us of how diverse the people who typed Google.com into their browsers were and how we needed to think inclusively about design if we wanted to capture the next billion internet users.

It was also a unique experience to get to spend so much time with Marissa. Back then she was about thirty-two, and it was evident she was beginning to feel the age gap between herself and the APM classes. We were members of the sixth class, and the experience of hanging out with us was different than it had been for her six years earlier on the first APM trip. She complained that earlier classes had tacked on hours more of partying after their twelve-to-fourteen-hour workdays, and how we were missing out on a critical part of the trip. We didn't point out that the APM trip back then lasted less than a week. It's one thing to run on four hours of sleep every night for five or six days, but it's a totally different thing to do it for seventeen or eighteen days. I could read between the lines. It seemed she also missed the days when it was still at least somewhat appropriate for her to go out partying with the APMs. Silicon Valley celebrities are still human beings. Spending so much time with Marissa would give me empathy for Mark Zuckerberg. It enabled me to better appreciate what it must have been like for her, or for him, to step onto a start-up rocket ship when you're barely twenty and stay on that rocket, riding it through all you have to ride it through until it's worth tens or hundreds of billions of dollars.

The trip unfurled ever westward as the days passed and we crept around the world. The totality of the experience felt like the fulfillment of my wildest dreams when I was sitting in Iowa feeling alone and isolated. Each night we had elaborate dinners with either start-ups or people from the local Google office. The camaraderie was like nothing I had ever experienced. We were bonded by the hardship of fourteen-hour day after fourteen-hour day, the natural hiccups of travel that are nostalgic in retro-spect, and the cultural experiences given us in each city.

One night, the staff brought out dessert after we wrapped a Thai cook-ing class in Bangkok and I still savor the joy I experienced as my friend David had a peak life experience. He stared at his plate, jaw literally hang-ing open, flabbergasted because he had been seeking a mangosteen, "the queen of fruits," for years, ever since he had read an article featuring inter-views with Michelin-star chefs about what they would choose as their last meal. The chef for one of the most renowned Thai restaurants in the world said that all he would have is a bowl of mangosteens, a small white fruit nestled in a thick purple skin. My friend had searched for a mangosteen for years, but back then the US Department of Agriculture had banned importation. Because they can carry a fungus that can endanger agricul-ture, he had never gotten to taste one. Now mangosteens sat in front of each of us. I can't imagine they would have tasted so sweet if we hadn't heard him pour out his heart about how his quest was finally fulfilled. Pleasure shared is pleasure multiplied.

CHAPTER 5

Silicon Valley

A mind that is stretched by a new experience
can never go back to its old dimensions.
                    —Oliver Wendell Holmes, Sr., *The Deacon's Masterpiece*

In the fall of 2007, after the round-the-world trip with my cohort of associate product managers and Marissa Mayer, I ended up going to work at Google Books. Because I followed my passions to a less popular corner of Google, I picked up some specialized experience that would later open the door to work at Pinterest and Facebook.

All APMs rotated to new teams after their first year at Google, a policy that struck me as an excellent management strategy. If we remained on the first team to which we were assigned, the people around us would likely treat us as if we were still flailing Nooglers. Google Books wasn't my idea—Marissa was the one who suggested it to me, during the pre-rotation counseling session she did with each of the APMs. Beyond sharing her greater insight into what opportunities existed and letting us leverage her experience with what drove successful APM rotations, she also helped keep us from running lemming-like at the same handful of coveted positions that in reality might not be the best match for our individual skills or goals. During that meeting with Marissa in her glass-walled office cube, the moment she suggested the Books team, I felt a rush. It seemed a magnificent idea; why hadn't I considered it?

In the interview with my soon-to-be Books boss, Adam Smith, he began with the obvious question: "Why do you want to join Google

Books?" My reasons gushed out of me as if I'd been waiting for this moment all my life and hadn't known it. I explained I had been a Cold War Studies minor in college; I talked about how when the Berlin Wall had fallen and the Soviet Union dissolved, a handful of academics rushed into the Soviet Union to scan as much of the Soviet archives to microfiche as possible. Their intuition about the need for an independent archive available to the public was spot-on. After one too many pieces of scholarship were published that bruised the egos of those who still looked back on the days of the Soviet Union fondly, Russia clamped down on much of the Soviet-era documentation and locked it away in government-controlled archives. I shared with Adam my dream that one day Fortuna would turn the wheel of history and the opportunity to finish the job would open. I wanted to be ready to scan the whole archive so that the full truth (from at least the Soviet side) of the Cold War would be known.

I was overjoyed when I found out I had gotten the job. It would take many months for me to realize that it was actually a job few people would want, and almost certainly a job that just about any senior product manager would have counseled me against taking. Many if not most Googlers viewed the Books team as unsexy work — at best a nonprofit hiding in a for-profit. It was culturally marginalized within the company.

Google Books arose directly out of an insanely ambitious vision of Larry and Sergey. Google was going to scan every book ever written in every language in the world and make them accessible to everyone. It was core to the Google Search mission of organizing the world's information and making it universally accessible and useful. Books would be a massive extension of the reach of their search engine. Some of the first mechanical engineering teams at Google had built elaborate book-scanning machines that lowered the cost of scanning books by more than an order of magnitude, allowing Google to scan tens of millions of books "economically" for the first time in history. I put "economically" in quotes because Google invested a third of a billion dollars or more to accomplish this feat. When I joined Google Books, the company was not yet three years post-IPO and still in the golden era when its core business raked in 75 percent profit margins. This was one of those rare periods when there was virtually zero pressure to make financially "reasonable" decisions.

Google converted bound books, sometimes centuries old, into clean, crisp, flat rectangles that could easily be processed by a computer or viewed on a screen. The procedure was so simple that all it required was a human to sit and flip the pages one by one using a foot treadle like those on an old-fashioned sewing machine to cue the cameras to shoot each fresh set of pages. This near miracle had been accomplished by using innovations like lasers to scan and model the three-dimensional curvature of the pages as the book lay open under a pair of high-resolution DSLR cameras shooting across each other, one facing each page. Google's software engineers used this 3D model to "dewarp" the curved pages of an open book and project the photographs in tidy, consistent, flat images. This technique was combined with the world's best optical character recognition (OCR), the translation of images of characters into computer code—the information that allows one to render text on a website—and in the widest array of languages ever produced. This was the quintessential Google golden age project no one else would or could have done at that time.

I would later contemplate the difference between what Facebook chose to invest in during its windfall years versus what Google chose to build during its own period of unprecedented abundance. Consider the contrast between Google books and Facebook's Free Basics. Free Basics was marketed as a plan to bring internet access to the poorest people around the world. If you used Facebook to access the internet, your data was free, but if you wanted to use anything on the open web, outside of Facebook, it was going to cost you. Facebook said they did this because they cared about providing internet access to people.

The reality was that Facebook wanted a monopoly; they cared about getting their own network to a level of penetration where no competitor could challenge them. It was ingrained in Facebook that they came to prominence because they had killed off Myspace and Friendster, and Facebook was always looking over its shoulder for the next Facebook killer that might be coming for them. Free Basics set up a wall that would be difficult for any company to scale.

I remember the first time I toured the book-scanning factory in Mountain View, a site that Google would not allow to be photographed because it invariably conjured up comparisons to sweatshops: row after row of book-scanning stations manned by workers who rarely lasted even months because the tedium of doing nothing other than flipping pages eight hours a day was so soul-crushing. The sound of people tapping their feet to propel their machines evoked an image of a Victorian sweatshop producing endless piles of garments.

I was rotating onto Google Books alongside another APM to backfill for Rupert, who was rotating off. Rupert was one of the few hires with a master's degree in library science, and he was brought in to help launch Google Books. He was also the rare APM not to have rotated until two and half years into his tenure at Google. In one of my first conversations with Rupert I said how excited I was to be working on Google Books because I had spent countless hours in research libraries as an adolescent. I gushed about the sound of silence when one is surrounded by stacks of books, and the distinctive scent of old volumes. Adam looked me in the eye and with the mild scowl I would come to know as his default expression. "Smell..." he began, his pause pregnant with the disapproval he reeked of, "...is a sign of *decay*." Fortunately, I only had to interact with him for a few weeks as the other APM and I onboarded to replace him.

That was the environment I entered at the start of my second rotation at Google. Because I was the first APM to hold the title of second/junior APM on the team, I was something of a floater. My job was to do whatever needed to be done. Initially that meant to handle the internationalization of Google Books. Unlike my peer APM's eight- or ten-person team, I was assigned a single engineer, Antoun, to take on all the tasks required to expand Google Books from maybe fifteen languages to forty as part of the Forty Languages Initiative Google kicked off that year.

Part of why Google has been so successful and is the default search engine for the world was their focus on making their products available in a much broader set of languages much earlier than other tech companies. *Internationalization*, also known as "i18n" ("internationalization" begins with an *i,* has 18 intervening characters, and ends with an *n*), and *localization* (L10n, though usually with a lower-case L) are the processes involved in making products available in new languages or adapting them to be locally relevant as well as accessible in the local language.

Google was far ahead of Facebook in this regard. My Facebook disclosures would reveal Facebook's many toxic practices and the impacts that it knew about and hid. The issue I would spend the most energy emphasizing is an underlying factor that is not intuitively obvious — and therefore, perhaps, especially insidious. Facebook told the world that artificial intelligence would save us from the perceived dangers of Facebook algorithms by selectively removing "dangerous" content. Yet Facebook concealed a critical fact: Focusing on "dangerous" content required translating and rebuilding those AI censorship systems language by language — a process that they had no plan to scale economically to an extent that they could keep people safe in the poorest and most vulnerable places in the world. In other words, although Facebook took steps to make its English-language content safe, the company would not and could not take those same steps for most other languages around the world.

Much of the work I initially did for Google Books was basic project management and not product management — chasing approvals from Legal and asking translators to translate the strings (words and phrases) used in the user-facing interfaces of the website (the front end) into new languages. But some of my work was more significant, like deciding what criteria a language would need to meet to be eligible for Google Books to launch it as a fully supported language. In consultation with the Books Search Quality team, we settled on the rule that Google had to have scanned 10,000 books in a given language before we could launch Google Books in that tongue. For context, a small Iowa library might have from 1,000 to 3,000 books. The University of Iowa has 5,000,000 books and a central library in a major US city might have several million. I recognize that 10,000 books may not sound like that many, but I learned in the process of preparing for our launch that in 2007 most languages had extremely little content available on the internet. Ten thousand books represented a *noticeable* fraction of all content available on the internet in, say, Bengali or in Norwegian.

Another criterion for launch that we established was that the OCR developed by Google had to meet a certain level of quality. Meaning, there could only be so many errors per thousand words, to ensure that users had high-quality search results. I particularly enjoyed this part of the project because it was my first experience getting to dig into the nature and nuance of writing systems—the family of characters that makes up a written language—and how much beauty exists in the diversity of alphabets (characters depicting units of sound) and syllabaries (characters depicting syllables) across the world. Computer scientists aren't typically offered the opportunity to relish the pleasure of learning about cultural details like this.

Internationalizing Google Books meant that for each of these new languages, Google would now blend results from Google Books into the results of any user's web search, and because of the lack of content online in most languages, Google Books would significantly enrich Google.com for those languages. I remember how satisfying it felt the day we turned on Google Books for a significant portion of the Google Top 40 languages (the most popular languages among internet users). It felt like proof of the nobility of our calling. We were organizing the world's information and making it universally accessible and useful. We were unlocking information for those who hadn't had access to it. I was really enjoying my new team and our mission. It seemed an example of technology making the world a better and more connected place.

Our engineering director of Google Books was Dan Clancy, who would also prove to be a vital mentor. Dan had previously assigned an engineer to build a Data Cube, a system that would collect data that could help the team understand how people used Google Books. This was in the early days of data science. We didn't have any data scientists on our team, and despite much nagging Dan couldn't get any engineers to actually use the Data Cube he commissioned.

He tasked me with figuring out how to make the best use of the Cube and to give him a dashboard to understand how Google Books was used. I spent a number of nights slamming my head against the wall as I taught myself SQL—the language used to request data from the Cube. The user interface for the Cube was one only a software engineer could love—a white page with a box outlined with a thin black border and a Submit button. Initially, I put one query after another into the box and downloaded files that I would import over and over again into a spreadsheet. It was a highly repetitive task that screamed to be solved with computer programming.

As an analog electrical engineer, I had learned to program in languages that were not normally used at Google. Sensing an opportunity to learn a more relevant way to code, I scanned the list of languages that already had packages (programs that facilitate a task or tasks) written for accessing the Data Cube and decided this was the moment to teach myself Python. I had written a small amount of code in it a few years before, so I knew just enough to think I knew more than I did. If I had not programmed in Python before, I probably would have read through a tutorial and avoided the hilarity and embarrassment that ensued.

Very quickly I ran into a challenge: I needed to use the "math" module to do enough calculations to provide Dan with a useful dashboard. Back then, Python wouldn't let you use fractions to get precise answers. If you divided 4 by 5 you would get 0, not 80 percent; dividing 5 by 4 would give you one, not one and two tenths. You had to add a math module to Python to get better answers. I started searching for a way to include the math library in my program, a process called importing. I tried Googling [how to import math module Python]. I asked colleagues, but no luck. I didn't understand why this question was so hard to answer. I soon learned I couldn't get help because the answer was so simple that the people I was asking probably couldn't imagine that I would ask such an obvious question.

During this time period at Google, I was often working late into the evenings trying to catch up from my late start. One night I poked my head out of my little office and prayed someone else was still working. Google Books was a team that very senior engineers joined because they wanted to give back. It had a higher fraction of senior engineers than any other large team at Google, and it was not a team that often worked past 5:30. I got super lucky, because sitting in one of the guest cubes was Matthew Gray, a new engineer on the Google Books Boston team. Matthew had played a pivotal role in the early web, putting up one of the first hundred web servers in history (www.mit.edu) and writing the first robot (a computer program, a bot) that crawled (wandered and tracked) the web. He used to tell me about how back then the web was so young (and small), that with a little help to avoid crawling infinite dynamically generated pages (like calendars), his crawl would complete even though it ran on a single machine. It was a time when it was still possible for a bot to touch every visible page in the web, and stop because it was finished. I felt like I was reliving history alongside him when he would recount the excitement he felt the day he realized the internet had started growing fast enough that his crawler could no longer keep up with it.

I walked over to him, introduced myself, and sheepishly explained that I had a very basic question that I was having trouble solving on my own: "How do I import the math module in Python?"

Without showing any reaction that might have made me feel any more embarrassed than I already was, he said calmly, "Import math." I went back to my office and flew through my code for the rest of the night. The kindness Matthew showed in this moment was the foundation for what would become a multiyear collaboration that changed the course of both the Google Books team and the larger Google Search org itself.

Just a few weeks later I found myself in a situation where I was a cyber-sleuth hunting for a "bad guy." My engineering director, Dan, wanted me to figure out who was scraping (downloading) the books off Google Books. Large numbers of IP addresses (the numbers that identify individual computers on the internet) were suddenly hitting the daily 1,000-PDF file-download limit for Google Books. Almost as quickly as Google was scanning books, someone was downloading as many as they could, and trying to get around Google's defenses to protect them.

I had an engineer help me learn how to access Google's raw logs, which recorded how people accessed the Google Books website, and started analyzing which IP addresses were downloading the PDFs. There were many unique addresses that were maxing out their download quotas, but it looked like they were all coming from the same area of the internet. Internet addresses back then followed a format called IP4, and were made up of four groups of numbers ranging from 0 to 256. For example, 192.168.0.1. All the addresses that were maxing out their download quota had the same first two digits (18). I walked over to the small office where Dan and my boss, the product management director, sat and presented them with what struck me as a curious pattern. Dan took one look at my screen and the clusters of IP addresses and said, "It's Aaron Swartz."

This seemed a magical feat to me. How could lists of numbers point so definitively to a single name? The internet was originally built to connect the networks of elite universities. Although the system could handle nearly 4.3 billion IP addresses, back in the mid 2000s there were still a significant number of "range holders" who controlled large blocks of IP real estate. These were such veterans of the birth of the internet as Hewlett-Packard, Stanford, IBM, Xerox, Bell Labs, and MIT. Dan immediately recognized that the "18" meant MIT.

For Dan, suspect number one became Open Library, an initiative of the Internet Archive, an online digital library project. And covert Open Library data access meant Aaron Swartz. He and Open Library had been critical of Google Books, and Swartz had developed a reputation for being less than observant of copyright laws, a reputation that would lead to his tragic death by suicide a few years later, when the federal government cornered him for similar behavior to what I had unearthed on Google Books and threatened him with up to thirty-five years, restitution, forfeiture, and a fine of up to $1 million. His prosecution was government grandstanding to warn off others who might violate copyright, and his death would make him a martyr. It wouldn't be the first time Open Library had been copying the books Google had paid to scan and stripping off the watermark. The issue for Google wasn't so much the downloading as Open Library letting Google pay for the labor of scanning the books (while drawing the ire of copyright holders), before Open Library swooped in to take credit for the output.

The question the Google Books team now faced was: How do we handle this? Dan and the team decided that twenty-one-year-old-punk Aaron maybe would respond better to me, a "kid" closer to him in age, simply asking him to stop, rather than senior engineers like Dan or, god forbid, getting the Google Legal team involved. I sent out an email later that day asking Aaron if we could have a quick call. He agreed.

On the phone with Aaron, I explained to him how I had been analyzing the traffic on Google Books and it was clear that someone at MIT was downloading huge quantities of books. I was hoping he would stop. He admitted he was the one and said he'd stop. Whether he actually did stop or just hid his activity better, I'll never know, but it was the start of a friendship between the two of us that would ultimately improve access to digital books. I had caught Aaron red-handed, and he respected that. Aaron was distinctly important to my early career because he was the first person who treated me as a technical peer. He didn't have a large window into my capabilities, but he knew I had caught him, and he always treated me with respect, and I was grateful for that.

Back in 2007 and 2008, it felt impossible to make any changes at Google Books. Google had scanned millions of books that we couldn't make available because the company was in the middle of one of the most expensive class-action lawsuits in history—the Google Books Class Action Lawsuit (catchy name, I know). Authors, agents, and publishers all argued that Google was illegally scanning their books without their permission and had sued Google for $3 billion, the proposed damages calculated using a per-book cost-and-profit-loss formula. Some of the issues were unclear— the mere act of scanning, for example, was not necessarily illegal (Google contended building a search engine was a "transformational" work, and therefore protected), but letting people have access to complete works that were under copyright, without compensation, clearly was. Google and the plaintiffs had been negotiating for years.

I hated it because it let people inside Google say, "Let's wait on that new feature. We don't want to spook the plaintiffs." If people had known it was going to take years and years to reach a deal, we might have released more features as we developed them. In the end, the final settlement left Google right where it had been at the beginning—Google could allow users to search its complete books database, but readers could view only limited sections of books that were under copyright.

That's where Aaron and Open Library came in. Open Library was a nonprofit project and no one was gunning to sue them—at least at the time. (Perhaps in a symbol of their success, more recently publishers have indeed begun to sue Open Library.) Aaron had a punk-rock vibe of *We'll just take the data if they won't give it to us. Information wants to be free.* The first gift he gave me was when Open Library launched a distributed scraping plan to find out which books on Google Books were available, in which countries, and at which "viewabilities."

Books remain under copyright protection, a system for controlling who can make copies of a work, in many countries for seventy years after the death of the author. When a work falls out of copyright, it becomes part of the public domain—anyone can reproduce it for free. Yet Google had scanned books both under and out of copyright, and had obtained licenses to make some of those copyright-protected books available in some countries but not others. Open Library released a plug-in for the popular web browser Firefox that allowed tens of thousands of people to lend their computers to figure out which books had been scanned, which were viewable, and by whom. Aaron's Open Library plug-in forced Google to react. Google had been repeatedly asked for this information, and had declined to provide it—now Open Library was going to take it. I didn't know it then, but discovering a system similar to Aaron's that was scraping Facebook over a decade later would prove to be a turning point on my whistleblowing journey.

Dan paired Boston-based Matthew and me on a project to launch the first application programming interface (API) for Google Books. This was the first time we would programmatically open up any part of Google Books to the rest of the world. The idea was to allow library patrons to search our entire database, and if they were able to view a book where they lived, they could receive a link to the book's online preview. Matthew and I dove in and soon had a product that was ready to ship. For years people had refused to link to Google books from places like library card catalogs because there was no guarantee a user could read the book once they clicked through. Now that hurdle was gone. This was my first experience walking through the product development journey from start to finish for a new product. I proudly informed our marketing manager, Melanie, we were ready to launch.

Melanie looked over the blog post I'd written to announce the release and gave a slight shrug. "How long would it take you to have some integrations [i.e., actual users]? It's ho-hum right now. What if we waited two weeks? What do you think you could do by then?" I looked at her with utter frustration. I had felt so proud of what Matthew and I had accomplished and I wanted it to be done and out. But Melanie was doing her job, and doing it well. She was there to help us tell the story of our product, and in order to tell the story of a developer API well, you need customers to act as proof of your interface's utility.

Melanie's feedback that day in late February 2008 prompted one of the favorite experiences of my career. I set out to see how many libraries we could get to integrate the API, starting with a couple of librarians famous for being coders who "rolled their own" library catalogs. Because these librarians wrote their catalogs from scratch and personally controlled the code that ran them, two small catalogs were able to integrate the preview API within a day. This let me lean on all the bigger catalogs with the line, "It's really not that hard. The Ann Arbor public library was able to do it and have it live in less than twenty-four hours."

The small local library catalogs could launch a new feature in a day because they were accountable to no one. If the small library had an engineer who was confident in their own code, and if they wanted it to go live, it was going to go live. Launching a feature in a larger platform, however, requires more than writing code—it requires internal checks and processes for things like stability (Will this crash everything else?), design reviews (Does this user interface make sense to someone who has never seen it before? Does it look good? Does it use colors that are on brand?), legal, and other sign-offs. Still, demonstrating that it could go live quickly made the hardest step—being willing to change—seem plausible. And that's what it took to be able to count fourteen integration partners on launch day. The library catalogs of the world could now point to Google Books and allow patrons to preview a sample of any book. Matthew and I shipped our product, and it was instantly in use. I knew in no small part I had Aaron Swartz to thank for my launch. It wouldn't be the last time Open Library would step in to sweep away naysayers who were preventing us from pushing Google Books forward.

Finally, I felt like I was hitting my stride at Google. Throwing myself into work was satisfying, the problems so messy and nuanced it was easy to get fully engrossed with them. Around the time of the launch, I was introduced to the process of Search Quality, the part of the company responsible for making search results better over time. Despite Search being essential to Google's brand and business, Search Quality had not been steadily building a community of product managers who deeply knew the space, unlike almost every other focus area in the company. Product managers would regularly appear in Search as their first posting inside Google, and a year later, as soon as they were allowed to, they would rotate off.

Part of the reason for this was that Search Quality was sufficiently complicated that someone with only a few months' experience was likely to suggest strategies that had been tried five or six times before. The search-quality engineers kept a detailed wiki they could point you to when you thought you had a brilliant idea. Chances were, they'd heard it before and maybe even in a better form, and unfortunately the idea was almost certainly not worthwhile. If you were an ambitious product manager straight out of business school, if you rotated onward to Ads or somewhere within the universe of Google Apps, it was more likely you would end up in a role where you could clearly demonstrate and feel your value. As a result, no critical mass of product managers stayed in Search.

It was Dan Clancy who was responsible for the Books Search Quality team. When I appeared on the team, Dan decided that rather than wait for a product manager who likely would never be allocated, he would just make a new Search Quality PM out of me. He would sit me down multiple times a week for lectures on information retrieval theory and the process of working on search products. Because of his investment in me, I was able to be part of the founding wave of algorithmic product managers in the industry, which ended up setting me on my path to everything that would happen at Facebook.

The person who would have been my manager if I had been a Search Quality product manager on the Web Search team was Lucas, and I would regularly walk across the street from my building and join him in his office in the late afternoon. I came to be friends with his office mates, who were some of the original generation of engineers who had built Google and now were senior members of the Search team. This was the era when few people except statisticians and quantitative user researchers were expected to understand how to work with data, and Lucas and I would spend evenings in the office trying to figure out how to extract information from the logs, the raw records of how Google's services are accessed. Almost no product managers in the company scoured the streams of user search behavior like we did, and it was as if we had our own private Russian novel. Most people have no one in their lives that they're as honest with as they are with Google, and after peeking through this rawest and most human lens into the human condition, I came away changed. Once you see people's unfiltered thoughts, you have one of two choices. You can throw your hands up and say, "This is too much," or you can come to accept that all humans have flaws and love them the way they are. I knew there were few people in the world with a telescope into the human condition like ours. We were in on the birth of a field, and it was exhilarating.

Meanwhile, as I was dancing with the poetry that was search queries, my manager, Noah, was growing progressively more anxious about our tracking metrics. Something was profoundly wrong.

Noah didn't really get how the nuts and bolts of search worked—that was something he delegated to Dan—but he could see on the Google Books tracking dashboard how much less often books were showing up in the Google.com search results. On any given day, half as many books were appearing there, and each day's decline clearly formed a steep curve heading toward zero. Noah was more than irritated. His performance was evaluated in part by how much traffic Google Books received, so every "missing book" from a search was a slight cut into his evaluation, and thus his bonus. Every week or two he would raise the issue again, but the core Search Quality group had recently reduced my engineering team from ten-plus people to only one person, Jacob. Each time I raised the issue with Jacob, he came away from his investigations more perplexed. I would learn much later that he was not alone in being unable to pinpoint the goblin lurking in the code that was hindering the performance of the system. Dan had asked some of the former engineers from Books Search Quality, engineers much more experienced and senior to Jacob, to try to figure out what was going wrong and they too had come up empty.

One day Matthew, my colleague from Boston, and another member of his team, Chris, were visiting Mountain View. I shared with them my problem. Matthew was itching to prove that his Boston Books team could handle more than Mountain View's table scraps. He said he and Chris were there to help. In no time, word somehow got out about our plan to work together, and an engineering director on Search Quality promptly scheduled a meeting with me to shut us down. He warned us that Google had learned the hard way that trying to spin up new search offices failed far more often than it succeeded — there was far too much accumulated but unwritten knowledge that went into Search Quality.

But it wasn't his head count to direct — Matthew and Chris reported to Google Books and not Search Quality. I asked him directly, "Are you going to allocate engineers to fix the books blending slide?" (That was Googlespeak for the fact that books weren't showing up alongside web results on Google.com anymore.) We stared at each other. We both knew the answer. The core Search team thought books were of questionable value — all the information in them was by definition "old" as far as they were concerned. That's why they had gutted the Search Quality engineering team in the first place.

So our pirate band leaned in, and before leaving Mountain View, I scheduled Matthew and Chris for a circuit of meetings to help them cram into their heads whatever unwritten knowledge they could find on how Google's Search stack operated. Back east, they pulled in another Books Boston engineer who had formerly worked on Search Quality in New York. In a little less than three days, they solved the problem. Book search traffic doubled over night.

I was dying to know what the cause was, and so they explained. Let's start with a fact every Googler knows: Google does not run on a single giant computer, some all-knowing brain in a warehouse somewhere. Early Google engineers had pioneered the process of using cheap "commodity" hardware (processors, hard drives, and so on) and stitching them together in a way that it didn't matter that at any given time some fraction of the hardware might be broken. This was in contrast to the industry's best practice at the time of using a small number of extremely expensive high-performance machines designed to never go down. Google's search stack was designed to be run on a very different "parallel" infrastructure. Google relied on many smaller computers connected together, each running their own copy of the Google Web Server (a GWS instance, pronounced *gahwis*).

Everything was designed so that after you searched once, the next time you searched you might hit a different computer and not have any idea the switch had taken place.

This system was very efficient to maintain — you could take individual computers offline and update them and stand them back up, and traffic would flow between them uninterrupted. But there were glitches. Or so we learned. The three musketeers out in Boston found the rare error case. It turned out that a large distributed system could hide pernicious problems. Matthew and his small team had discovered that at some point, months back, a software update had started rolling out through the Google Web Servers, first to a few computers and then on to more and more each week. If it had been a simple A-to-B switch rolled out to every server at once, the cause would have been obvious — our traffic would have plummeted on the day the change was made. Books were there one day, no books the next day. The reason our traffic had been slowly declining for months was that this had been a riskier change, and the Web Search engineers had rolled it out incrementally to more computers over time, in the process disguising that the change even took place. A tiny bug had been introduced that blocked the updated GWS instances from knowing how to interact with our Books servers, effectively closing off the channel that brought books to the Google Search page.

At Facebook, I think part of why many of the worst impacts were missed was because they were gradual. Misinformation farms slowly ramped up over years. The feedback cycle between algorithms rewarding extreme content, and content producers picking up on those changes and creating more extreme posts, took months or years. The reality is that companies governed by metrics and experimentation have trouble identifying slow causal patterns. They're used to seeing changes that cleanly demonstrate that X caused Y. Most of the worst impacts of Facebook's product or algorithmic design choices happened step by step, with each step adding thin layers of sediment to the mountain of harm. The infection was slow — and then it was everywhere.

In solving the Books search problem, what Matthew and the Boston crew demonstrated is how vitally important it is to be mindful of this possibility. If you're not open to entertaining the right questions, if you don't allow for the possibility of slow-moving changes, they can sink a business division, and maybe even the entire business, over time. Or, in the case of Facebook, they can sink a democracy.

———————

That winter was the beginning of my last days at Google. I had successfully extended my deferral by a year, but if I didn't pull the trigger and actually leave Google for Harvard Business School, I would have to reapply to attend. I felt incredibly conflicted. I loved the people I worked with on Book Search, particularly the Google Boston team, but it was also clear to me I would always be working uphill to convince people that Google Books mattered. If I was going to go to the trouble of switching context and rolling the dice for new people I would be surrounding myself with after moving on, now seemed as good a time as any to at least see what HBS was like.

That winter had been particularly hard on me. Since I was a child, I had a higher baseline of stomach problems than the average person, but not enough to actually motivate any primary care doctor to try to resolve them. Something about the grind of the previous couple of years had kicked my health challenges into higher gear. At first I lost ten pounds. With the irrational invincibility of youth, I just appreciated losing the weight. (Note: If your weight starts plummeting with no explanation, don't follow my example!) By a few months later I had lost thirty pounds, and my body was very clearly falling apart. I avoided eating. My face was suddenly covered in a patchwork of acne that seemed like it would never heal. I went repeatedly to the doctor, and despite multiple rounds of testing, I still felt like I was stumbling blind.

Lucas was still my Search Quality mentor and had the advantage as my not-actual manager of being able to ask personal questions about my health. He noticed that something was profoundly wrong and kept nagging me to return to the doctor after each successive round of inconclusive tests. One of the struggles of being really ill is you lack the energy to advocate for yourself the way you need to when the medical establishment shrugs. After the third round of testing that produced more questions than answers, he started poring over my test results, reading up on various conditions and how they might manifest in my symptoms.

After studying page after page of results and utilizing Google as only someone who had shaped it could do, he came across my "high normal" score for celiac disease. "Are you aware celiac has, like, fifteen symptoms, of which you have at least ten?" he said. "Why don't you try giving up gluten?" He had invested so much time in trying to debug what was happening to my body, the least I could do was try. And sure enough, Lucas was right. My doctors hadn't figured it out, but he did, with the help of Dr. Google.

Google's Mountain View campus is one of the best places in the world to explore gluten-free eating. Within three days I stopped running to the bathroom constantly. Within two weeks my skin evolved from a patchwork of acne scars to perfectly clear. When I visited the Google Boston office a few weeks later, one of my engineers was so shocked at my transformation he enthusiastically remarked, "Frances! You look so much better!" I chose to interpret this as a compliment. I felt like Lazarus.

And just in time. I needed my strength to pack up my apartment and move cross-country. Google had been an emergency parachute—I still feel like it chose me, rather than the other way around, when they took a chance on me that no one else would. If I wanted more than what my life was now, it was time to spread my wings and try something new. Off I flew.

CHAPTER 6

Veritas

Physical and psychological adversity shape us.
Our challenges give us insights and experiences
that only we have had, and . . . we need to not only accept,
but also embrace and see as strengths.
While we may not have chosen to include them as concepts of
　　ourselves,
they are there. And what more can we do but own them?

—Amy Cuddy, *Presence*

Even if you've never set foot on the Harvard University campus in Cambridge, Massachusetts, you might have notions of what it's like. Whatever those images are in your mind's eye, they're probably accurate. Harvard is America's oldest institution of higher education, dating back to 1636, and for the most part it looks like it. According to the American Association of Landscape Architects, the main grassy Yard at the heart of the campus is one of the oldest continuously used built landscapes in the United States, and the centuries-old, stately buildings that ring it have come to define what colleges look like: neoclassical architecture attempting to evoke a feeling of established eternal age. Walking across the Yard on one of the many gray fall days in Boston, if somehow you could blind yourself to the clothes and hairstyles of the students you pass, you could easily imagine yourself strolling along these paths in any year since 1850.

The portion of the university's two hundred acres that comprises the Harvard Business School campus is much more contemporary. I spent most of my time in two main buildings: the Spangler Student Center, built in 2001 and named after a billionaire alum; and Aldrich Hall, built in 1953, financed by a gift from John D. Rockefeller Jr. and named in honor of his father-in-law, and refreshed with extensive renovations in 2004. Most of my classes were held in Aldrich, in large lecture rooms that resembled those of the United Nations, both in terms of the midcentury-mod interior design and the stacked, stadium-style horseshoes of seats filled with an international mix of students. I didn't live on campus; through a Google connection, I rented a small in-law cottage with a slate roof that gave off a fairy-tale vibe. Between classes I hung out in Spangler, where extensive lounges around the dining hall let students spread out with study groups or socialize.

I entered HBS in the fall of 2009, part of the first wave of students to matriculate after the 2008 financial crisis. As the market declined, head count declined with it, and HBS took advantage of the "high number of high-quality applicants" to "increas[e] the number of people admitted to the MBA program." I ended up being among the 30 percent of "nontraditional students" — not an alumna of a financial service firm or consultancy — who were required to show up two weeks before the start of the academic year in order to take a primer on fundamentals of Finance and Accounting. That two-week period also gave us time to get settled and generate friendships among our group before the McKinsey types and Wall Streeters arrived — precisely the kind of people the vast majority of Googlers would have avoided in San Francisco.

I had arranged to attend Harvard via the Google Fellowship Program. Google would reimburse my tuition in exchange for me committing to return to the company for two years. I mention this for three reasons: First, it's an example of the investment Google was willing to make in their employees. Second, almost no Googlers at the time took advantage of the program; I was among the first employees to do so, maybe the very first. Finally, although Google was willing to finance a substantial portion of my Harvard education, many people at Google discouraged me from going to HBS.

Coworkers whom I deeply respected had warned me away from attending. Some of the discouragement was gentle. Over coffee, an HBS alum who was the product manager for the company's revolutionary Google Wave told me I should go join a start-up instead. An engineering director told me I should go get a computer science doctorate, and that going for an MBA was a "waste of a mind." I had pushed ahead despite the discouragement and now sat a continent away from Mountain View because I cared intensely about organizational design and the practice of people management. During my first few days at Harvard, I thought about that feedback and wondered whether they had been right.

I was not well suited to sitting still in class for eighty minutes at a time, just listening. On my first day of class, I abruptly awoke to the fact that I had never before been part of a traditional educational setting. I had gone to a Montessori preschool, a highly student-directed and free-form environment. I attended a Montessori-like elementary school. Olin was Montessori-like in its project-based education. And Google was about as close to Montessori for adults as you can get—an environment that actively values internal motivation and creativity. I was cast in roles where I could chase things I was passionate about. My job at Google was kinetic. Now I had to will myself to sit still and focus on the conversation ping-ponging between my classmates and the professor. We were coached at the start of the semester not to attempt to speak in every class, and told that if we kept raising our hands, we would not be called on. Staying quiet and sitting still have never been among my strong suits.

After the remainder of the class arrived, they gathered all of us in a large auditorium on the first day of the semester. A presenter paced the stage and gave what endures for me as one of the most valuable lectures I would hear during my two years at Harvard. He warned that it would be easy to get swept up in the false-wisdom groupthink of mid-to-late twentysomething type A's plowing toward financial success. He instructed us to think more expansively. Mindful of our egos, he said that we were all very talented and that we should remember that when we were ten years old none of us probably dreamed of being a McKinsey partner. (Though to be fair, I suspect at least some of my classmates did indeed have such dreams at that age, but I digress.) Regardless, hundreds of HBS alumni every year go into consulting and investment banking. The presenter encouraged us to remember our childhood dreams and go achieve those.

This was a theme, a reflection, that HBS tried to reinforce throughout the program. When you pursue what you're passionate about, you can end up in what's called a Blue Ocean. In business, according to this particular framing, you can either go elbow to elbow with competitors trying to meet existing demand (competing as just another shark in a roiling Red Ocean full of sharks and blood) *or* you can go create something entirely new. Maybe you'll inspire novel consumer demand. Maybe you'll invent a whole new space. Cirque du Soleil was held up as an example of entrepreneurs inventing a new category. Its vision of what a circus could be was so different from the legacy circus companies of Ringling Bros. and Barnum & Bailey that Cirque du Soleil created a whole new category — a Blue Ocean — that it easily dominated on its own.

HBS was acknowledging that we students who were driven enough to make it to Harvard Business School, who would *want* to attend such a school, might be limiting ourselves both professionally and personally by chasing externally motivated goals. Instructors encouraged us to consider and reconsider our priorities and what really mattered. If we truly followed our passions, we wouldn't climb the most obvious and most crowded ladder to success, we'd find one of our own. We had the ability to create a path to our own success — our own joy — but we had to be brave enough to choose to.

That was only one way HBS was clearly trying to shape not just our practical education, but our education as people, because they understood it was vital to our long-term success. In a class called LEAD (Leadership and Organizational Behavior), we reviewed profiles of the Class of 1976. We read and discussed autobiographical essays written by members of this class at the time of their ten- and twenty-year reunions. Across the board, at ten years out the alumni were doing well. And yet at twenty years out, they were not. Their partners had left them, their children were distant, some had flamed out of their professional pursuits for self-inflicted reasons. They drove away the people who were the engines of their success because they were so blinkered by a laser-like focus on that success that they couldn't navigate their own lives.

Our professor told us these bios were not cherry-picked; they were representative of the outcomes of that class. We needed humility to recognize that we had to cultivate whole lives and care for the people around us if we wanted to reach our dreams.

———————

Because HBS has more than nine hundred students in each graduating class, they divide the class into ten sections to give it a more intimate feel and facilitate closer relationships. Even today, you attend all of your classes, from about 8:40 until 2:30, with your section of some ninety students. You share a series of coordinated social events with them over the course of your first year on campus—following a template supplied by the Student Senate and the Department of Student Life—to help instill coherence in the "HBS experience." Having attended a start-up college with no traditions, I welcomed this. It was like I was getting a second chance at a "traditional" college experience.

Over the next two years, I learned as much, if not more, from the in-class discussions as I did from the texts or the lectures. The takeaways, at least my takeaways, from those discussions would very much come to mind when I was considering what, if anything, I should do at Facebook. One particular case-study discussion from my Political Economy class stands out in my memory. Earlier in the semester we had studied wealth inequality in Chile, and throughout the session many of us, myself included, had questioned whether this was really such a big deal. According to the quantitative measures we were being taught, the United States had worse wealth inequality, and no one seemed to be raising alarms about it here.

Then the last class of the semester rolled around and the hammer dropped. Most classes at HBS are taught using the case method. Instead of buying textbooks at the start of the semester, we would work from stapled booklets—one for each case. Sometimes cases were given to us at the start of the semester as part of stacks of paper so large that twine would be tied around them to make them manageable, the sheer number of individual sheets accumulating into small towers in my bedroom. Other times cases would appear in our physical mailboxes only days before the class in which they would be discussed—today's case was one of those surprise ones they had held back intentionally so we could not anticipate it. We had spent the semester turning our eyes abroad to how politics and economics impacted business around the world. Now we would bring it home and discuss wealth inequality in the United States and how it contributed to social instability.

I don't remember all the details of the discussion, but the punch line was clear. The professor said something along the lines of: "You have a *really sweet deal*. Each of you is walking out of here with a Harvard MBA. Your life is going to be comparatively smooth sailing. Do. Not. Mess. It. Up." And by "It" the instructor didn't mean just our lives, but the world. We were still only a few years on the other side of the Enron scandal, wherein a bunch of Harvard MBAs had let their greed and guile lead them into sufficiently illegal spaces that they had made and then lost billions of dollars. Many innocent victims lost their life savings, and some of the largest energy markets in the United States were terrorized with rolling blackouts. The message of the class was clear: Social stability should not be taken for granted; even if we weren't going to be governed by a higher creed, we had a personal interest in doing the Right Thing.

Almost every classroom included someone who had real-world experience in the subject matter of each case study, which would infuse the discussion with specialized and current insight, and sometimes in the most unexpected and illuminating of ways. This was center stage one day when our Ethics professor asked my section whether we thought sending executives to jail was effective. A classmate, and member of one of the *chaebol* dynasties, raised his hand and gave an answer no one saw coming. Unlike in the United States, South Korea's economy was largely run by the *chaebol*, a handful of massive, mostly family-run business conglomerates, and those different economic circumstances influenced different business expectations.

"I don't think sending executives to jail is very effective," he said. The class stirred. The student shrugged. "When my uncle went to jail, we just made him the head of the company again when he got out." Heads snapped to attention. This was why HBS spent so much energy trying to bring together a diverse set of people. Expectations are not uniform. To truly be successful, you need to understand your context and be mindful of the fact that there may be context you can't possibly fathom until you do the research. Even then there are some things, some perspectives, some motives you can never see coming — so acknowledge that possibility, too.

Part of why Facebook executives were so blindsided when ethnic violence emerged in Myanmar, and by their role in it, was that they did not invest proactively in channels for lifesaving information. They assumed their products could not be misused in deadly ways, and they valued efficiency over collaboratively working with the civil society groups who had boots on the ground. The reality is we cannot expect companies to do an adequate job at this without outside input and oversight — no one intentionally works on things they think are dangerous — because we all naturally develop blinders when left alone.

I soon came to the conclusion that I was incredibly fortunate I had persisted through the discouragement to actually attend HBS. Yes, I could have constructed a syllabus and forced my way through it, but I don't think I would have ever chosen the different interweaving branches of the tree HBS had spent decades building. I grew not only from the information, but because the environment and my peers forced me to grow. I came away from the first semester deeply grateful for what I was learning. I had no idea yet that my time at HBS would provide a scaffold that helped me rebuild my life later on when unforeseeable circumstances would leave me a husk of my former self, and would go on to inform my decisions about what to do at Facebook.

For all its strengths, HBS was not an unadulteratedly positive experience. My number one regret is that I lived off campus. I wish I had lived in the dorms because it would have created more opportunities to meet people and develop deeper friendships. The little in-law unit I lived in was set in the backyard of an engineering director at Google. It was picturesque with its slate roof, and I loved it, but it made it too easy to retreat into isolation.

At first it was a sanctuary because I wasn't fitting in with my section. I had come from a place, Google, which magazines regularly proclaimed was the best place to work in the world. What limited business experience I had was rooted in a fairy-tale setting where there were 75 percent profit margins on revenue that grew by 20 or 30 percent every year. This gave me a lack of empathy for people who had to make compromises within tight and rapidly fluctuating constraints. I had the arrogance of someone whose corporate cult espoused the motto "Don't be evil."

HBS is an engineered experience to the point where we had assigned seats in our classes, and it wasn't just so that it was easier for the professors to take attendance. HBS wanted to prevent hypercompetitive students from jockeying for the best seats every day. It also wanted to ensure that the children of billionaires, who had been shipped to HBS with hopes they would be inspired to be gainful contributors to society, didn't just get high or drunk and spend their time in class sleeping in the back row.

My assigned seat was at the end of the front and lowest row in the class-room. It was tied for being the most visible seat in the room — everyone could see me. By chance, a group of men who mostly came from invest-ment banking happened to have seats together in the rear top row of our section. From that perch, the "Sky Deck," they literally looked down on me, audibly scoffing every time I would raise my hand and speak. Day after day. No matter what I said. I could not move, I could not hide; my assigned seat held me in front of their withering gaze.

The climate outside the classroom also began to affect my mood. After years of California sunshine, Boston's gray fall and winter skies were get-ting to me. One late fall day, after a class with the scoffers in the back, I stepped out into the gray fog and chilling mist and felt overwhelmed. I found what I thought was a private area on the grassy common facing the river in the back of campus, sat down, and sobbed. I couldn't fight the con-stant strain. A friend from my section happened to walk by on the sidewalk maybe 100 feet away and saw me. "Frances, are you okay?" she called, clearly concerned. I waved her off. "It's okay, I'm fine," I said. I pulled myself together and got up and moved along. Nothing to see here.

My greatest wish, sitting here writing this today, is that I could go back and tell her, "No, I'm not okay." If I had, she probably would have come over, she might have even tried to find me some support in our sec-tion. I was locked in a pattern I had repeated before in my life — what I really needed was help, but I didn't know how to ask. It would take years, and a near-death experience, for me to escape the cell I had built for myself, but I didn't know that yet.

"Hitting bottom" is a misnomer. You can be certain you're done falling, but you can always surprise yourself by going deeper. You have to earn your right to hit bottom by changing your direction. Looking back, I wish I had confronted those guys in the back of the class, not with hostility but rather one on one, maybe invited them out to coffee and talk. Asked them directly, but politely, to back off. Who knows? It might have surprised them enough to work. We all find ourselves in circumstances where we think we cannot meaningfully change the dynamics of our situation. The only way forward is to find even the smallest actions like that and try to do something different.

But back there on the lawn I was not the person I am now. I didn't see a path to meaningfully change my environment. I didn't try to alter my circumstances because I didn't have enough faith in the ability of others to change, or maybe I didn't value myself enough to demand I be treated better. But — *but* — I could change my frame of mind. I saw there was fundamentally a choice I could make: I could let them get to me and "win," or I could ignore them and keep going. They were doing what they were doing because my sheer existence annoyed them. The best revenge is living a good life. Soon after that day on the lawn, I changed my mindset, and that made their behavior easier to bear. I now had a purpose.

I didn't fold, and I made it to the spring semester without leaving HBS, as one of my section mates had. When I returned from winter break, what I witnessed made me reevaluate how I had approached the conflict just a few months before. A group of women in my section, marshaled by our section president Kamala Avila-Salmon, decided they didn't like living in the toxic environment of which the bullying directed at me was only a small slice.

Part of why I wonder if I could have stopped the scoffers sooner if I had just taken them out for coffee or recruited other help from within the section was that when those women came together and announced "This is over," it stopped. Individuals can change. Circumstances can change. Even whole groups of people can change. If I had just believed that change was possible, if I had just tried solving problems with others instead of alone, who knows how different that year might have been.

This moment marked the time when the Class of 2011 Section B finished moving through the first three stages of team formation we had learned about that fall: We had formed, we had stormed, we had normed, and now we could perform. My on-campus experience improved dramatically.

Throughout my time at Harvard, I worked part-time at the Boston office of Google as an intern. Many engineers before me had part-time appointments at Google while they pursued their master's or PhD. I was the first product manager who had gotten such a part-time appointment. It had taken my team going all the way up to the engineering vice-president for the East Coast to get approval, but it was viewed as worthwhile at a minimum to keep me around for institutional knowledge. I was the only remaining continuous thread in Book Search over the past few years.

During my first year at Harvard, Google was a life raft, a fact I deeply appreciated but that probably kept me from growing my way out of the challenges I faced at HBS. I was embedded within my familiar team on Google Books. Book Search still did not have a data scientist (or "quantitative user researcher," as they were called back then) to do logs analysis, and so I was tasked to build out an analysis library for Book Search. My team was full of people I had known for years and gave me the threads of social connection I needed to keep going. I would pay extra attention in class each day, looking for the pithy anecdote I could regurgitate when a Google colleague would ask, "What did you learn in class today?" Even as I was gaining a more grounded and nuanced perspective on Google, I still felt profoundly tied to and grateful for it.

It was then I also made significant technical progress at Google. When I gave Matthew my first pass at the logs analysis library, his response was charitable but clear. He led me to a conference room so no one would have to hear his critique. "Frances, your code . . . runs," he said with a pause, "but we need to cover some fundamentals of modern software engineering." Thanks to Matthew I learned how to structure my code for reusability, testability, and readability. I graduated from coding like an electrical engineer to coding like a software engineer. I couldn't help but compare and contrast Matthew's demeanor to the scoffers in my section. With the right support and large amounts of time to practice, it was in this environment that I really fell in love with programming and the actual building of products.

For about two years I had been intrigued by one question: How do we help less savvy users access information in books? Books are complicated objects, and we take for granted the skills that allow us to scan a table of contents or an index and find the answer we seek within the book or, perhaps more importantly, to realize we had asked the wrong question in the first place.

> In the next section I mention many things about Google that will likely seem foreign in the near future if "large language models" continue to grow more capable and prominent. As I finalize this manuscript, Microsoft has announced the first "conversational" search experiences, in which an artificial intelligence can actively chat with the searcher and guide their process of inquiry. So, if you prefer, treat this as a trip back to a vanished world.

Up until then, Google had treated a book as if it were a website. Websites have many pages. Books have many pages. The internet has many websites. Google has many scanned books. Google is extremely powerful if you want a discrete answer to a specific question; it will comb all the websites it can access and find you specific web pages that have information similar to the words you put in the search box. Hopefully that information will answer your question, though Google makes no promises it will. It is a search engine, not an answer engine. Google Book Search had similar limitations, only within the domain of books Google had scanned. You couldn't ask questions like "What should I know about Historical Figure X?" or "What can I learn from Book Y?" No matter your search query, you could only find lists of individual book pages in response to your queries, and they would throw them at you in some order only partially understood even by Googlers themselves. Much like Google, Google Books was very useful for people who already had a clear and extremely targeted question, or who understood how to get to an extremely targeted question quickly. Google Books was unmatched for people who knew how to use it to do more complicated research. Unfortunately, people who excel at such research comprise only a tiny sliver of people who use the internet. Just as Paul Muret had taken the complicated thought processes of website traffic analysis and made a user interface that helped unlock this way of thinking, I wanted to unlock for more people the many ways you can use a book.

While some might argue that even today Google designs for people like themselves, in the past this was even more pronounced. I remember sitting down with a senior engineering leader I respected and showing them a histogram of how many times people did a search on Google in one day. Sixty percent of users searched only once or twice. "Should we be designing so intently for the 1 percent of users who search the most?" I asked.

What happened next shows you how different the world is today in terms of how we use data to drive design. That search engineer looked at my graph in disbelief and asked, "How can this be right? How can someone only search once? Don't they learn something and search again?"

Corporate culture has been described as what was successful in the past. For most of Google's lifespan, it had taken so long to bring novel technologies to market that Google basically designed products for their own needs. "By the time this product comes to market, the world will have caught up," the thinking went. I like to say data science is about empathy building, which might seem counterintuitive, given many people's perception of statistics and numbers. Looking at a graph forces us to imagine people different from ourselves, but only if you believe in data-grounded decision making. The Googler I was speaking to couldn't fathom my graph because the average Googler searched twenty or twenty-five times a day. As you can imagine, it was hard for most Google search engineers to ask, "What kind of experience would be best for the typical user?"

Many years later, think pieces would begin circulating on Gen Z's use of TikTok to answer many kinds of questions. The consensus from the young people who were interviewed for these articles was that they liked how complicated information was synthesized on TikTok—a clear video was more convenient than having to stitch together a cohesive opinion, fact by fact, by yourself using Google.

You never want to give interns mission-critical work. It's stressful for them, and it's stressful for you, their mentor/manager. During those HBS years I was an intern, and my concern with people who struggled to use books effectively was considered sufficiently noncritical that it was a perfect problem for me to work on during the summer between my two years of HBS. I was excited to get to spend a couple of months building a product for more average users. I began from the insight that books are not web pages — they are designed products brimming with intentionality. The information you read in the first chapter of this book was placed there intentionally, just as the information in this chapter has been placed here intentionally. The structuring of the information is so important that multiple people, including an editor or copy editor, might have weighed in on whether an anecdote should have gone in this place or that to convey information most effectively to the reader. If the same "chapter keywords" — words and phrases that Google deemed to be significant in each chapter — reappeared across multiple chapters, one could assume they were more important than chapter keywords that appeared in a single chapter.

My user interface used this insight to find the most important topics in a book and then showed how those terms appeared as the chapters progressed in a condensed enough manner that trends became visible. I'll give you an example about how my interface helped me to discover something unexpected about Alan Turing, widely considered to be the father of modern computing, whose code-breaking machine built during World War II is believed to have shortened the war by years and saved millions of lives. I thought I knew who this man was. He was up there in the pantheon of computer science with Ada Lovelace and Charles Babbage. Yet when I went to spot-check the quality of the algorithm for prioritizing keywords contained using a biography about him, "cyanide-laced apple" was one of the top topics.

This had to be a bug, a flaw, in my code.

I ran my eyes down the interface and because the information was structured the same way it was laid out in the book — small summaries for each chapter, listed in chapter order—a coherent story jumped out at me. Turing's favorite movie was Disney's *Snow White and the Seven Dwarfs,* a movie so revolutionary in its time that it briefly held the record of highest-grossing "talkie" film. He was fascinated with the transformation of the Wicked Queen and the ambiguity of the poisoned apple. I knew that after he was charged and convicted under antihomosexuality laws dating back to the Victorian era, he had committed suicide. But I definitely did not know that the means he used was a cyanide-laced apple. The biographer had left a trail of bread crumbs about that apple woven throughout the story so the coroner's inquest at the end would make sense. My interface and algorithms valued books as their own special thing; *books are not just websites on paper,* and this magical moment was the result.

For someone reading this book, it may come as a surprise to learn that most people (and probably fewer every year) know how to use a book's table of contents or the index to expand the set of questions they might want to ask. Google Books probably would have taken you to the right part of that biography if you knew enough to phrase your query as [alan turing suicide], but even then it likely wouldn't have found the *best page* to help you understand *why* he died this way. Best-case scenario, you'd have to figure out for yourself that there was a collection of pages where it was discussed at length, if such a cluster even existed.

To me, Google Books was the *Star Trek: Enterprise* of card catalogs—we could rocket you to the right page in the largest library available online, but only if you knew how to ask your question precisely enough. My Turing moment made me feel like I had built a bicycle for the mind, something that would carry (more) people to unexpected places. It was my first experience of using technology to unfold what could be known.

You might ask, "Frances, if your interface was so good, why can't I use it today?" That summer I learned what happens when new technology rubs up against old laws, and the fact that laws and regulations are not free—you may lose experiences or products you value in exchange for the good the regulation will hopefully bring. You will never get a magical moment like the one I stumbled upon when you open Google Books, because when we worked through the launch review process, Google discovered that my algorithm might violate the "moral rights" of the creators of those books. In Europe, there is a concept called the "integrity of a work," which means that if you are a creator, you have the right to prevent your work from being distorted. You and your heirs sometimes retain those rights in perpetuity. The example I was given was to imagine you were an architect and you designed a bridge over a river. If a hundred years later the city wanted to add lights to the bridge, your heirs could invoke your moral rights and block that option if they felt it was in conflict with the architect's original vision.

I get when people urge caution in passing laws regarding social media. It's easy to look at problems we see flowing through our social streams and demand they be addressed immediately, but it's hard to see what's lost when more friction is added to the system. For me it isn't a hypothetical objection about hypothetical costs—I loved the interface I had built, and I felt it could make books more accessible to tens or hundreds of millions of people. But it couldn't be launched.

By my second year at HBS, I was in the home stretch, and I now only took electives on topics I chose. Not surprisingly, my last few classes were among my favorites.

This might seem unlikely, but history classes were among the hardest to get into at HBS when I was there (may that be a hint to the HBS administration on how to expand course listings...), and I had to trade away much of my schedule for lower-demand classes to "buy" my seats in the ones I took. I don't regret it. One of them, The History of Managerial Capitalism, still comes back to me across the decade-plus since I took it. We began the semester at the dawn of modern capitalism, when basically all large ventures were managed by their founders or the founders' heirs. Over time, we learned that the world developed in such a way that owners (usually shareholders) and the people who manage/run enterprises became separate groups. Very few Fortune 500 companies have their founders as their CEOs. It made me appreciate that the world we live in today is not immutable. In fact, many things we take for granted are decades, not centuries, old.

I'm not a business historian, so I won't go too deep, but here's a relatively simple example. Even what we think of as a corporation is a fairly new invention. When you see reports in the news on how a company is faring, you'll often see something like "Meta's new metaverse division lost $2.8 billion last quarter." A statement like this assumes that companies have divisions that are responsible for controlling their costs and driving up their revenues — that they can, and we expect them to, track and report independent profit and loss amounts for each division. Remarkably, this configuration has only existed since the 1920s, when the DuPont corporation reflected on a paradox. They had just exited the First World War, a time when they should have been printing money, but despite having an ideal market for their explosives, they had run a deficit. DuPont realized that unless each division was forced to be transparent about their expenses and revenue, the company as a whole could end up losing money. Division leaders did not sufficiently scrutinize their operations for efficiency unless they were accountable for those decisions. These kinds of insights — the idea that the world as we know it is actually quite new — gave me the freedom to think more expansively about our ability to drive change.

Probably the most important thing I learned at HBS was that the world doesn't operate on two-year time horizons. In that history class, and in every case study over the two years of my MBA that featured the narrative of a single big company, I learned that large, meaningful things take ten or more years to build or complete. This was a revelation to someone coming out of Silicon Valley.

When I left for business school in 2009, the tech industry had very short-term horizons. I had spent my first three years at Google surrounded by Stanford graduates who obsessed about crazy-fast returns. We had all been raised on stories of the internet wild west in the 1990s. The canonical rocket ship was Netscape, which IPOed sixteen months after it was founded and was valued at $2.9 billion within days of open trading. Even if people couldn't expect to IPO that fast anymore, the way money is raised in Silicon Valley trains people to imagine only things that can be accomplished within two to three years. This thinking permeated companies like Google, where promising projects like Wave — similar in many ways to the wildly successful app Slack, which would come down the pipeline just a few years later—were killed because they didn't take off fast enough.

They say happiness is just the difference between expectations and outcomes. Choosing to believe that anything truly interesting takes ten years to accomplish has been a core part of why I can keep pushing for transparency and accountability for social media. Significant change requires shifting behavior and overcoming significant inertia, all of which takes time.

Shifting my mindset was liberating because it gave me a sense of control. I was and am a cyclist. I'm familiar with the joy of the grind. Happiness is not the end, it's the process. Having a long-term perspective was essential after I came out as the Facebook whistleblower. After the initial explosion of attention, I began to receive a surprising number of panicked messages: What was I going to do to keep the story in the public's eye? Was the moment for change lost?

Meaningful change, real change, happens slowly. The thing that most often blocks us from real change is the fear that we're not being successful enough fast enough.

The other great mindset shift I experienced that spring was due to a class called Power and Influence, taught by Amy Cuddy. While this might seem like a cliché of a business school class, it felt in many ways like the primer I had never had someone explain to me before. Its goal was to help us understand and identify power and influence dynamics and learn to use them as effective tools for analyzing our surroundings and achieving our goals. The class was easily 75 percent female, and the majority of the men who joined us were people of color. Just as my seminar on the History of the Family at Wellesley had attracted undergrads who wanted to study family dynamics, my Power and Influence classmates seemed to illustrate that I wasn't the only one trying to make the effort I exerted and the returns I received match those of my peers at Google.

I still remember my final paper for the class. We were asked to look ahead to the next couple of years and the challenges we were going to tackle and write a plan for how we would use what we had learned in the course to reach our goals. Cuddy would later leave Harvard, but she unquestionably changed the course of my career because she helped me feel that I had agency to navigate my return to Google. There were things about me I could not change—I was still going to be among the radical minority of women at Google—but there were many things I could control about how I interacted with others and recruited support. And I was going to need that support.

CHAPTER 7

So Close

Healing severe or chronic pain...includes transforming our relationship to the pain, and, ultimately, it is about transforming our relationship to who we are and to life.

—Sarah Anne Shockley, *The Pain Companion*

In the darkness of predawn, Connor and I stood atop a dormant volcano. It was late summer of 2011, and we were on Maui, on the island's highest peak, Haleakala, the "house of the sun." Haleakala provides a view of the sunrise that is legendary. Mark Twain described it as the "sublimest spectacle" he ever witnessed. We were among a small crowd of tourists. The air was crisp, making Connor's hand feel warm in mine. A seemingly endless gauze of clouds stretched across the sky beneath us.

As dawn broke and the sky began transitioning from dark blue to purple there was a collective gasp from practically everyone gathered on the peak. A symphony of colors slowly and somehow all at once filled the world around us, fiery crimson mixed with orange and gold hues. Then came the crescendo—the incandescent star emerged over the horizon, emanating a glow, a light, unlike anything I had ever seen. I felt gratitude for the so many divine interventions that had made this moment possible.

The months after I left business school had been a series of childhood dreams made real. I was part of the most exciting team at Google, which is to say, Google+. I was building my dream home with friends. I was married. I was independent and successful in all of the ways that then mattered most to me. Even my presence on the mountain was a symbol of my success—I hadn't technically qualified for the Google+ celebration trip because of the date I had joined the team, but because I'd delivered so much value that summer, I was one of a handful of exceptions who had been flown out to Maui on Google's dime along with the rest of the team to toast our successful launch. As I stood on that house of the sun, feeling the joy of being there and being there with Connor, I indeed felt on top of the world.

Connor and I had met years before on the Google Books team, and during my time interning at Google during business school we had become much more than friends. We had eloped to a beach on Zanzibar over the winter holidays my second year. Now we were on another tropical island, and the buzz I felt as the sunset peaked resonated with echoes of the joys from that previous paradise.

Only a few minutes later, however, almost as quickly as those colors in that sky had changed, my bliss became something else, and I was reminded of the weakest link in our ability to make a difference in the world. Although the sun had cleared the rim of the peak, we weren't done with our ceiling-of-the-world experience. We started up a trail to a viewing plat-form at a higher elevation, hoping to eke out a few more minutes of joy. Less than two hundred feet later, I suddenly had trouble breathing. I waved for Connor and the friend who was with us to go ahead; I told them I was fine, just needed a minute. It was a short, maybe 300-foot, walk up the spiraling path that wrapped around the rise. It might take me a little lon-ger to get up there, but I'd quickly catch up. Right? I told myself that the cold had caused my asthma to flare up—dry cold weather is the only thing that has ever triggered it even mildly—and I just needed to walk slower. I watched them pull away from me and disappear around a bend in the path as it rose toward the true peak. I thought I was fine. I thought I could do it alone. I was wrong, both on that trail and in life in general, but I hadn't learned that yet. I made it through that hike, but it would take another two years for the full wake-up call to arrive.

The business school class I'm happiest I took was Authentic Leadership Development. I initially regarded it as a touchy-feely blowoff course and only signed up for it because I wanted a class schedule that would let me travel to the Bay Area when I wanted and to have time to work on projects at Google. The premise of Authentic Leadership Development was rooted in a concept born of interviews with world-class CEOs. It's easy to believe CEOs are where they are because they took an escalator straight to the top, but every one of them confessed they had encountered a moment when they faced what amounted to their crucible, a crisis or series of crises that almost derailed their lives. Each of them attributed their respective successful careers, in large part, to the fact that they had faced what seemed like overwhelming adversity and they had figured out a way to overcome and emerge from that crucible stronger and wiser. ALD tried to help us anticipate that this could happen to us as well, and to prepare us with tools to make sure we soared instead of sank when trouble would inevitably arise.

The years immediately following Harvard Business School were my crucible. That's why, of all the chapters in this book, this one has been the most difficult to write. But the experiences I lived through would inform the choices I would make at Facebook and made me the person I am today. Without this period of my life, I might never even have worked at Facebook, let alone blown the whistle on what was taking place there.

One of the core concepts of Authentic Leadership Development involved recognizing and accepting our whole selves, including the parts we viewed as the imperfect parts of ourselves and our pasts. Lying to ourselves and others by hiding those perceived limitations drains away our energy and limits our potential. All but the most charmed among us, not just CEOs, will confront a crucible moment. That's life. Perhaps something in my crucible experience will resonate with you and help you feel less alone in yours.

Up on Haleakala, at the break of day, I was absolutely alone. I had gotten far enough from the parking lot that I could not see it, and my companions were somewhere ahead of me, when my knees buckled beneath me. I *could not breathe.* This wasn't some mild chest tightness like I had felt once or twice before. I was down on the ground, head between my hands, dark spots dancing in my peripheral vision. For the first, but not the last, time, I needed help but couldn't call out to those who might have given it to me.

It was a moment that foreshadowed all that was to come in the next two years. A severe, life-threatening constellation of illnesses would creep slowly into my life in a way that was difficult to diagnose. I would find myself divorced, unemployed, broke, and quite literally fighting to survive. This series of crucible moments forced me to learn the skills that led Facebook to hire me, the same skills that would drive me to make the choices I would make there. The challenges I would face as I lived the journey of my whistleblower experience would not seem nearly as daunting as they otherwise might have. As I would respond in the French Assembly when asked what had given me the courage to act, when perhaps 200,000 employees who had come before me at Facebook had not, I truthfully paraphrased Camus: "Once you conquer your fear of death, anything is possible."

———————

The Google I returned to was a Google on the move. Like many other large tech companies including Meta/Facebook, Google was best able to execute when it had a clear and articulable enemy to battle against. I returned to the Mountain View campus in May 2011 when that opposing party was Facebook. Fear of the social media giant was powering the engine of initiative inside Google, pushing the company to move into "social."

This wasn't the first productive crisis I'd witnessed. Google had swung through one of these phases before I left for grad school when it realized it was falling behind with the mobile internet. I had watched many mobile web versions of Google's products, including a mobile Book Reader I product managed, launch rapidly. Countless reviews and speed bumps were streamlined or skipped. Later the urgency built to the point of releasing a full-blown mobile operating system (Android) to compete with Apple's iOS. Red tape and, even more importantly, skepticism fell away. Throughout the company, there was an urgency to push forward fast.

Google relied on the open internet; its core business model was web search and the search-based advertising that appeared beside the results. Because the open web is free of walls, available to anyone, Google's bots, computer programs that download information, could cast a wide net. They spanned the vast majority of the accessible internet, collected hard to imagine quantities of data, and allowed Google to present the captured information in clean search results with clickable blue links. Facebook, on the other hand, was closed. Google couldn't search any of what was inside Facebook's walled garden, which was a growing, separate universe of data.

Google had a unique god's-eye view of the internet as a whole. Better than anyone else, they could see as they navigated the web who was creating content, and by monitoring the clicks on their search results (among other sources of data), what content was being consumed. The trend lines did not look good. People were spending progressively more time on Facebook, and the content creators had followed the eyeballs — people weren't investing in blogs or other forms of independent and open hosting at the same intensity as they had before.

Thus Google regarded Facebook as an existential threat, an aggregating black hole of content competing to be the default information operating system of the world. As a result, Google commissioned a team to develop Google+, tasking it with leading the company into "social." I angled hard to earn a place on the most buzzed-about team at the company (or at least the most buzzed-about since the previous Google social hub, the often controversial Google Buzz).

Google+ was a textbook example of what I saw advocated in the Change Management course I took at HBS. The most important takeaway was simple, laid out in case study after case study: It's hard for an individual to change, and when you scale that challenge up to a team or an entire organization, it becomes exponentially more difficult. One of the most important things leaders of an organization can do to increase the chance for successful change is to form a group that will drive the change and carry the banner of where the company is going, and then support that vanguard unwaveringly. Every organization has entrenched interests; after all, people have made lots of choices and sacrifices based on the *old* way of doing business. Unless you establish, elevate, and enable the vanguard, recruit more and more members to that new cause and push the laggards into their camp, you will not succeed at change. Years later at Facebook, the company's decision to abandon its Civic Integrity division would directly lead to my decision to blow the whistle. I had seen what it took to change Google — there was no way it would be easier at Facebook.

In 2011, Google+ was meant to be everything you loved about Google, with a little bit extra. That "plus" was supposed to be a new social layer, which is to say that Google's products would be more aware of human relationships and would facilitate collaboration and sharing. But to allow relationships to happen on your software, you must have consistent concepts of who you are (identity) and how you want to interact with others (permissions). A handful of Google products had tried to add social aspects to their experiences over the years, but a mismatch of different versions of "you" and different methods for indicating that you wanted others to see your data or communicate with you had grown up without any clear centralized corporate guidance. Google+ was intended to clean up the confusion and create a common framework for identity and connection across the entire suite of Google products.

You could tell that the Google leadership had gotten the Change Management talk from McKinsey or whichever consultants they brought in, because they went to great lengths to make us feel special. Google was 100 percent committed to making sure it won at social. Our team was located in the brand-new towers that had just been built (or bought and renovated?) on the campus. Even our proximity to Google founders Larry and Sergey, on the floor right beneath their new offices, broadcast how important they thought the project was. Google+ was at the very heart of the Googleplex. Having spent four years on Google Books, which was never well loved, it felt like I had arrived.

My project was among the critical components of Google+. A couple of the early social pioneer products at Google had come together years before to create a central Google Profile for each user, but this had never been essential to using Google. You probably never even realized you *had* a profile floating around the internet with your name and other personal details you entered when you first started using Google (RSS) Reader or some other product years before. A few other pioneer products had competitor profiles that were even more shambolic. As part of Google+'s imperial sweep across all of Google, the Google+ team had anointed the Google Profile victor, rebranded it the +Profile, and made it the canonical home for standardizing all personal data across the company. It sounded important, but as I would learn soon after I rotated onto the team, the integration work that was the core of its value had already been completed.

But just because it was largely complete didn't mean it didn't attract controversy. My first summer back at Google brought a directly related movement called Real Names Considered Harmful. In techland, this movement was part of something dubbed the Nymwars (a play on "pseud-onyms"), which was stoking controversy within Google. The leadership of Google+ believed that Facebook's real-names-only policy was a critical pillar in the success of the product because it drove greater accountability. Policies like real names only, however, can stigmatize and alienate people with nontraditional names. There are some cultures, for example, where people go by only one name (a mononym), but all Google "real names" had to have a first and last name. There are people with "made-up names" like "X Æ A-12," which their parents, and likely the holders of the name, would certainly consider real. A real-names-only policy would exclude them, and could also exclude people who can't use their own name online because to do so would compromise their safety (survivors of domestic abuse and stalking or, for that matter, whistleblowers). Protecting anonymity becomes even more important when we consider illiberal regimes. Software written presuming it will be used by people safely covered by the First Amendment is routinely exported and adopted in societies where people can be persecuted for minor offenses like making a joke about their ruler. Within Google, a group rose up to oppose real names only. Considered by some to be the first large-scale organizing effort within the company, Google employees demanded the real-names-only policy be rescinded. Ten percent of the engineering staff came together to sign a petition demanding as much.

Working on the +Profile, I was at the center of the controversy. As the product manager for the centralized repository of data about an individual, I also handled "identity." Almost daily I would get pressure from people who wanted the policy to be changed, but the leadership of Google+ stood firm. In some ways it's symptomatic of how superficial Google's strategy for social was. They surveyed the history of social communities and had correctly identified that Facebook's "clean, well-lit space" had been a huge driver in helping them win the "personal" social media market. But Google ignored how Facebook had initially rolled out — it had moved slowly, beginning with individual college campuses. People had formed tightly knit networks with real people they knew, who they'd found with, yes, their real names. But Google wanted Google+ to grow fast, so they basically gave up on that "personal" use case. If you were an early adopter, you saw content from the few other early adopters who were there, not the people you knew on a day-to-day basis by their real names.

This was my first exposure to large-scale corporate organizing and to how myriad agendas within a company can undermine an honest assessment of the facts, and as a result undermine objective attempts to arrive at a solution. This experience would give me empathy for the executives who surround Mark Zuckerberg at Facebook. In my role as the product manager for +Profile, I would read emails from coworkers, sometimes half a world away, pleading for the policy change, and I would think, *I agree with you, but there's no point rocking the boat, I know the Boss, and he's not going to change his mind. Relitigating this for the twentieth time isn't going to do anything.*

For those of you who have children, the next time you're facing challenges from a willful or complaining kid, remember that if they're being unpleasant, it means they feel safe enough to do so. That is a parenting accomplishment, even if it won't feel that way in the moment. Children who don't feel safe don't act up. When I arrived at Facebook years later, I would be struck by the weakness of petition drives compared to what I had seen at Google. Even on vital issues of geopolitical significance, you could only muster a couple hundred signatures. I know why I didn't sign—I didn't want a target on my back. As Mark Zuckerberg liked to say at company meetings, "Facebook is not a democracy."

The Nymwars also marked my first experience with the limitations of content moderation. The goal was ostensibly simple: find and remove accounts with "fake" names. To give you a sense of why this is hard, you could almost certainly do a "pretty good" first pass at taking down fake names by, say, taking a list of names from the United States Social Security Administration and checking whether users had a "valid" first and last name that appeared within that list. Then again, many people have names that aren't on such rosters. A nontrivial number of the most senior engineers at Google had online handles they had personally and professionally gone by for twenty to thirty-plus years. Were their names not "real" at that point? And of course, anyone can adopt a pseudonym that borrows someone else's "valid" first and last name.

Nymwars was also my first experience with the pitfalls of an English-speaking company serving a diverse, international user base. Among those who would write complaining about the real-names-only policy were people who would have to transliterate their names. English uses Latin characters (A, B, C) to form the alphabet of our written language, but not all languages do this. Russian, Ukrainian, and dozens of other languages use the Cyrillic non-Latin alphabet (А, Б, В, Г). Hebrew (א, ב, ג, ד), Arabic (ﺏ, ﺕ, ﺍ, ﺍ), and Ge'ez (ሀ, ለ, ሐ, መ) alphabets are also non-Latin. Converting the alphabets of other languages into a Latin-letter-based alphabet can result in many written variants for a single name because characters in different alphabets don't cleanly map the same sounds one to one. The last name of Russian composer Tchaikovsky, for example, has been spelled in seven different ways in the Latin alphabet. Inside the company itself, many engineers from abroad had names that were flagged by the content moderation systems as not matching what Google+ considered a real name.

Much like the content moderation teams I would later run into at Facebook, the team responsible for removing "fake" names also understood they didn't have enough moderators to adequately enforce the real-names policy effectively and evenly across the site. Instead, they reasonably chose to focus their attention in a way that made their lives easier. They tried to maintain a minimum standard of care across Google+, and then tried extra hard to monitor every comment thread on every public post from the most senior leaders of Google+ in order to minimize critical comments on their "failings" from those leaders. The Google+ team couldn't stop all uses of fake names, but they could at least ensure that the bosses wouldn't see any.

And sure enough, over time, the leadership of Google+ thought the real-names team was doing a great job because they never saw "fake" names. Part of why I have compassion for Mark Zuckerberg is that on Google+ I watched well-meaning people hide hard truths from our Big Boss and his senior team. Effective leaders have to make sure they have channels that allow them to hear bad news and reward people who bring it to them. If you don't constantly check whether the channels are open, they won't be open when you really need them.

Over the course of the summer and fall, I did what little I could to make the +Profile/identity on Google more effective for more users. I rolled out features like mononyms (the initial version of Google+ only supported people with two names), and I worked with the Google+ team to give users the ability to withhold listing their gender on their profile, which was another concern of many Googlers and the public. Women are often harassed online, and to protect themselves from strangers, some prefer to hide their gender. These might seem like small victories, but even they were hard won.

———————

My +Profile work did not occupy all of my time, so I went looking for opportunities. I wanted to parlay my search-quality experience to build out full Search on Google+. Google+ had launched with only Profile Search — which is essential if you want users to add their friends — but it was less obvious to the leadership that retrieving posts was also critical. Search at Google was "transactional" — it was all about how fast they could get you from your query to the one single thing you were looking for. They quite reasonably pointed out that people don't necessarily come to social networks for specific pieces of information (unless you're looking for your long-lost love or the latest celebrity you just saw a YouTube video about). I thought Search in a social context could and should be a different experience. I thought if we did it right, it could be a bridge from the early days of Google+, when few were online, to a future when hopefully everyone else had shown up.

The Google+ team also knew that when the platform launched, there would be a period when many users would be engaging in Google+ via "single-player mode." In the early days of any platform focused on connecting people, you will have people show up who don't yet have friends to interact with. All successful social platforms have ways to engage with the platform that are enjoyable for a single person to use by themselves in addition to those that require other people to be present to derive value from. Facebook is the great outlier in that its initial college rollout strategy really only worked because everyone joined at once — only later did they build out things like Pages and Groups to ensure that your feed is full even when your friends are absent. They grew slowly, and thus were able to create an expectation, at least for millennials (and for a time), that "everyone you knew" would be there.

Twitter in the early days had a sizable number of users who used the platform for microblogging. When I joined it in 2006, I went for years only using it to microblog my thoughts—I didn't know anyone on there, and I didn't care to read their posts. The social platform TikTok has always been loosely based on giving *you* the best possible content, no matter who it's from. As a result, you're not showing up every day to see your friends. Same with Pinterest—it was based on *you* following people whose taste you liked, whether or not your friends were present to see your boards.

These forms of social media worked in single-player mode because they were "broadcast" channels—content creators didn't necessarily care who saw their content, they just wanted as many people as possible to see it. This is why TikTok clones are possible, and present, within YouTube, Snapchat, and Instagram—people literally copy their videos and upload them onto new services to get more eyeballs and thus ad dollars.

Google+ wanted to be something more—it wanted to be the "Facebook killer." People were on Facebook because their friends were on Facebook. You didn't join Facebook because you were interested in content from influencers on the platform, you joined because all of your friends and family were there—you didn't want to miss out on life. You didn't create a post because you wanted a random someone (a rando) to see your baby photos, you created that post because you wanted *grandma* to see your baby photos. But the only way grandma was going to see those baby photos was if she was also using the service.

This "grandma and you" challenge was a tough nut for Google to crack. It would mean Google+ had to have good enough single-player-mode enticements to draw enough people to try the service and keep using it while their friends were steadily connecting with them, and then transition those users to expect to see personal social content from others. No other social network had ever gone from "broadcast" to "personal," but Google+ was going to do its best to try.

What if alongside your trickle of posts from people you knew, you could also get a feed of content on a particular subject that flowed and updated? I dug through some internal data sources and found proof that about a third of all the searches on Twitter were by people doing repetitive searches, sometimes for hours. The spring and summer of 2011 was when the Syrian civil war exploded, and the best news source in the world covering the conflict was Twitter. As Bashar al-Assad attacked protestors demanding democratic reforms, the protestors documented the violence in real time via Tweets uploaded through VPNs to the wider internet. We found people spending hours refreshing Twitter once or twice a minute, just to see the latest news.

Having a good search experience *could* be part of making single-player-mode enjoyable. We envisioned allowing people to save topics in the form of queries, hashtags, and locations in a Google+ sidebar so they could immediately connect with people around the world who shared a common interest, and they could have a global conversation about it. It was in some ways trying to create something analogous to Facebook groups without building the tooling that a formal "groups" product would have.

Even with the data in hand, my bosses weren't convinced, until I heard a rumor that the Web Search team had noticed the same glaring absence and was going to make a move and ask Larry Page if they could take over that part of Google+. They had already built the backend infrastructure for searching posts as part of the core Google Web Search Google+ integration — they just needed Larry's blessing to own the user experience.

I knew an opportunity when I saw it. I presented a very simple choice to the G+ leadership — give me a few engineers to build out a front-end search team or accept that Web Search was going to own it. Fear of Missing Out (FOMO) is the secret engine of Silicon Valley — ask any venture capitalist or successful start-up founder. Suddenly, after weeks of pitching and cajoling, the leadership team made engineers available. I now had a project I was excited about that I had birthed from nothing. It was a rush. Maybe that Power and Influence course had paid off.

That same summer, parts of my life launched that were equally thrilling. Connor had stayed behind in Boston for a month while I started my job at Google so he could handle the packing. I had drawn up a schedule months before graduation of packing a box a day, but he had insisted he'd rather take responsibility for the packing and handle it over a time horizon that he controlled. I had persuaded the Google relocation folks to allocate us two pickups in Boston, one for each of our homes there. When I showed up in Boston over the July Fourth weekend, I discovered that nowhere near enough packing had occurred at either home to be ready for the scheduled pickups.

Boston was in the midst of an oppressive heat wave when I arrived. The air conditioner in Connor's 100-year-old brownstone was pretty much worthless. At some point during our wild rush, Connor suggested we drive somewhere to get pizza. He picked a shop far enough away that we could sit in an air-conditioned car for at least an hour. It was a seemingly banal act on Connor's part, but it showed he was aware of how I was doing and that he was willing to care for me when I was struggling. Over the next three years I would come to appreciate his many acts of kindness, large and small, that were in fact acts of love for me. When I look back over everything that would soon unfold and lead to the end of our marriage, the thread I consistently see is that we did genuinely love each other.

Soon after Connor was out in California, we were in the midst of the Oakland home renovations. I had bought a property with two houses on it with two friends from college, and the main house needed a full gutting. We had bought it because we desired a home and community, and we were willing to do the work. In other parts of the country twenty-five years old may not be an unheard-of age to buy a home, but in the Bay Area it was uncommon. Even with the benefit of buying at the low of the 2010 real estate market, I had heavily leveraged myself financially. Looking back on my mid-twenties, I think the fact that we were all part of the first graduating class at Olin fueled our desire to build community this way. We were the "cold start" of the Olin community in the Bay Area, and we didn't have other (perhaps better?) playbooks for how to launch adult lives. We all wanted a sense of connection—sure, call it a sense of family—and the risk felt worth it. We proceeded to gut the house and design our new dream home. I remember the magic of laying out our draft floor plan at full scale in masking tape, on the floor of one of the auditoriums in Google's building 43 over a Saturday. We included rectangles of brown paper to represent countertops, cabinets, and appliances. Once everything was down, we could walk through our future home and imagine the life we wanted to live together, not to mention things as basic as where the toilet needed to be moved to.

That day marked the first time I sensed Connor's discomfort with "our life," though I didn't immediately see it for what it was at the time. Maybe in retrospect I couldn't imagine seeing it for what it was. Google+ had recently launched, and because a number of Googlers (like Connor and me) had been early active members, our follower counts swelled in the early weeks of Google+. We were nascent Google+ influencers. Instead of helping lay out the house or participate in decisions about whether we should move the stove a bit this way or that, while we cut and taped and darted around the auditorium Connor was absorbed in trying to document what we were doing for social media and livecasting what we were up to on Google+. I'm not sure when exactly Connor realized he didn't want to be crowded in a house, in a life, with me, but that was what was happening. I had bought the property with my friends when it was just the three of us, and I had committed that great error of youth: I hadn't imagined what my life would be like when I got older. Now I was only a couple of years older, and Connor and I would have to share a thirteen-foot-square bedroom—a frightening downgrade from his three-bedroom brownstone bachelor pad—until we could renovate the second house.

Of the seven and a half years I spent at Google, the last few weeks before the launch of Google+ Search are among my favorite work memories. We gathered daily for thirty-person midday "standups" in building 43. To get there, we'd take a ten-minute walk over the dry creek that separated us from the classic Googleplex. The air in the war room crackled with excitement. We were building a different vision of Search at Google, something entirely new. The adrenaline surge among all was palpable and infectious.

While our first version would return a stream of posts to the user after they searched, the new version would automatically update your feed every thirty seconds or so if you put it in "lean-back" mode—inspired by the people searching for "#Syria" over and over again for an hour at a time. It was a fresh window, or rather a portal, that allowed the world to flow to you, constantly updated without your needing to lift a finger.

The potential impact, the sense of community we were attempting to create, became evident when Steve Jobs died of pancreatic cancer. I remember walking through the locked doors that separated our high-security floors from the rest of the campus, and surveying the sea of sit-stand desks topped with thirty-inch monitors that stretched out in front of me. All across the Google+ open-floor plan people were clustered in groups of five or six around a display, silently watching the live updating stream of the search query "Steve Jobs." Everyone in the room was standing as if in vigil, and it seemed to me you could almost see the pain of the loss flowing to us live via the Google+ portal we had created. This was the first time I felt genuinely *moved* by the impact of something I had helped build and release into the world. For years to come I would regard this moment as a bittersweet peak of my career. Little did I know what was in store for me professionally and, more immediately, personally.

———————

That winter I received a diagnosis of neuropathy — pain you perceive not because of damage or actual physical pain, but rather because nerve damage causes errors in the signals that the nerves carry to your brain. Put simply, your mind misinterprets that noise as pain. Signals from my hands and feet had the farthest to travel, and thus ended up "feeling" the most like they were on fire because they accumulated the most noise by the time they reached my brain. Our minds also triage where to focus attention: a mild backache can disappear if you hit your thumb with a hammer by accident. Compared to the higher levels of noise/pain from my hands and feet, the neuropathic pain elsewhere was invisible.

This diagnosis helped me make sense of my past. I had felt relentless itching and tingling in my feet when I was nine or ten, and the only way I could make it bearable was by scalding my feet in the hottest water I could stand for as long as I could take it. When I was seven I started spraining my ankles over and over again, setting the stage for an ankle-foot brace later in college. Maybe the damage went back as far as when I was four, when I would refuse to wear shoes but would also not go barefoot, because while the shoes were incredibly uncomfortable, the grass underfoot was its own kind of horror as well.

When I was little I would often stay inside during recess at lunch because even being outdoors in the middle of the day in direct sunlight was unpleasant. Whenever my ankle was sprained, I would play on the computers in the computer lab. The irony is not lost on me that my early nerve damage might have been part of what set me up to become one of the few women at Google when I graduated from college. Often our greatest strengths are also our greatest weaknesses, but sometimes the opposite is true, too.

But why was I ill? I sat online reading about neuropathy's top causes, a who's who of terrifying maladies. Diabetes. Chemotherapy. HIV/AIDs. And then, bingo: autoimmune disorders. Suddenly I had a cause — my celiac disease hadn't just starved my skin of the nutrients it needed to heal when I had struggled with it a few years before, it had also been starving my nerves while agitating my immune system, possibly for decades. There was no cure except time, and strict compliance with a gluten-free diet could help — but I was already doing that, right?

I was prescribed gabapentin, a frontline drug for neuropathy that quiets the sensations in your nerves. It immediately hit me like a truck. I didn't know it then, but when you've been under high levels of systemic stress (like pain) for a long time, and some of that stress is removed, that new relief relaxes you. Think about being really cold, and cuddling up in a bed under a heated blanket; you relax more easily and fall asleep. You might not think of cold as systemic stress, but it is, and removing that stress can cause you to relax and let go.

For about six weeks I would sometimes find myself napping on couches around the Google office at random moments. I had not been sleeping well for years, and I had no idea why. We become incredibly good at tuning out the baseline of our lives, even if it comes at a cost to ourselves. But once I had caught up on sleep, it was as if I had never been awake before in my life. I began to experience a level of clarity that was utterly foreign to me. Everything was so vibrant at times it felt like I was in a video game of my own life, and in Technicolor.

As my doctor stepped up my neuropathy meds, weird but welcome things began happening. When I pulled up to a stoplight, the green lights were suddenly intensely bright — a decidedly different color than they had been before. Even my field of vision was expanding. One day I was sitting at a stoplight, hands at two and ten, waiting for the light to change. Only this time, when the red light flicked to green, for the first time I saw that there wasn't just one light straight ahead of me at the intersection that changed; the lights for crosswalks flanking the road to the left and right changed simultaneously as well. My vision was expanding, growing from a limited circle maybe the size of a beach ball held at arm's length to something approaching normal peripheral vision. I had been spending so much mental energy ignoring that I was in pain that it literally blinded me to anything not immediately in front of me. I had a new perspective on life, literally and figuratively.

Part of why I can have compassion for the Facebook employees who didn't blow the whistle before I did is that their situation and the state of the company worsened only gradually. It wasn't just that it was difficult to spot societal problems in the metrics, watching accumulating decisions getting made was like a frog slowly boiling. Few things are as painful as seeing a wrong get dismissed when you draw attention to it. Likely many learned just not to look or at least not see, rather than face that pain. I know from firsthand experience that our brains protect us from information that wears on our short-term survival no matter the long-term costs.

———————

Connor and I were trying. We took a trip to Southeast Asia that ended up being amazing and painful in turns. We began in Ho Chi Minh City and traveled via bus overland through Cambodia to Thailand, where we finished in Bangkok. We had gone to Cambodia to see one of the great wonders of the world, Angkor Wat, and eat my favorite cuisine, French-style Cambodian. While we may have intended it to be a restful, relaxation-centered vacation, it ended up providing a pivotal education that led directly to my blowing the whistle on Facebook.

We felt ourselves stretched too thin to organize such a trip for ourselves, so we placed ourselves in the hands of a package tour operator. Besides the ancient temples came tours of the most significant sites of the Cambodian genocide. I had visited Jewish history museums around the world, as well as memorials to the Holocaust like the Theresienstadt concentration camp while biking across Europe during the summer before business school. When those sites presented images or artifacts of the atrocities of anti-Semitism, they were always presented with context intended to help us prevent such tragedies from ever occurring again. Those places weren't temples to death. Compared to the history and description provided at those sites, however, the Cambodian genocide historical markers offered very little information. Stripped of context, all that remained was raw horror. That genocide took place from April 1975 to January 1979, during which time nearly a quarter of Cambodia's 7.8 million population was murdered. Unlike the Holocaust, the Rwanda genocide, or the Armenian genocide, in Cambodia it wasn't a specific ethnicity or religious group that was killed—it was everyone with a high school degree or more, anyone who spoke two languages, or anyone who was even moderately privileged.

I didn't set out to see the various sites dedicated to the Cambodian genocide. We had signed up for the tour to see Angkor Wat and for some activities along the way, like riding little motorized carts sixty miles an hour over the rails of the mothballed remnants of the colonial French rail system. But how can a person gaze up at a twenty-story tower of bones, the bodies of five thousand people, and not feel the need to ask "Why?" or "How could this happen?" I remember asking our tour guide at each genocide site those questions, and each time, she would change the subject or just walk away from me. As we sat in a restaurant in Bangkok our last night together, I figured I only had one more chance to ask. Finally she was willing to explain why. The tour agency had a policy that she didn't have to answer any questions about the genocide while she was in Cambodia because many of the people who had committed the genocide were still in power. That was why the memorials were so light on explanations. But now, in Thailand, the guide explained that she had once asked her grandmother why the genocide happened. All her grandmother had told her was, "You know, we never really liked them." Resentment is a debt that quietly accumulates until the time comes to pay the bill.

After I came home I dove into reading about the Cambodian genocide. This is my understanding of what happened, but I am not a specialist in ethnic violence studies, so please forgive any misunderstandings. Some of the facts are simple—the genocide had come on the heels of French decolonization. Most Cambodians had been restricted to living on the land they inhabited in an unfree state for generations, and simmering resentment was still attached to people with advantages secured when Cambodia was a colony, even if those advantages were in actuality quite modest. When Pol Pot began organizing his revolutionary Khmer Rouge, the revolutionaries warned that those who lived in the urban areas of Cambodia would become the new "colonial" overlords of the rural peasants. The only way to preserve the freedom so hard won from the French was to remove the threat the urbanites posed. The Khmer Rouge envisioned an agrarian-centered communism grounded in the "base people" who worked the land and would be the bulwark of the transformation. The urban "new people" would need to be forcibly "reeducated"—or eliminated.

One great irony of the Cambodian genocide is that scholars argue over whether the revolutionaries *actually understood* what an agrarian communist state would be like, or whether they were, in historian Michael Vickery's words, "petty-bourgeois radicals who had been overcome by peasantist romanticism." In other words, had they been swept away by misinformation about the joys of farm life to the point where they were willing to turn to violence to achieve it?

Flash forward to eight years later, when I was a member of Facebook's inaugural Civic Misinformation team, tasked with figuring out a strategy for dealing with waves of viral incendiary misinformation that were spreading in some of the most fragile places in the world (even in the United States). My peers understood we were dealing with "At Risk Countries"—the term Facebook used to describe states like Myanmar that were at risk of violence ignited by social media. I kept thinking of the Cambodian genocide and the twenty-story tower of bones. Both were situations in which neighbors killed neighbors because of a story that labeled people who had lived next to each other for generations as existential threats. How quickly could a story become the truth people perceived?

Shortly after I returned from Asia, it seemed my life began a slow spiral downward. The Google+ team had promised a grand vision that would be ready to launch by Google I/O, Google's annual corporate conference, that summer. Any project not viewed as critical-path to those goals was cut. All of the engineers on the Search team were reassigned, leaving me alone. I was relegated to a holding cubicle with three other product managers who had also suddenly lost their teams. We were told to figure out something new to build that would justify allocating new head count to Google+, otherwise we could wait for Google I/O to come and go. The reality was that there wasn't a chance any of the features or experiences we might pitch would ever be supported—nothing we could imagine would be more important than preventing Google+ from falling on its face when the summer came. For the next few months I clung to my job as other parts of my life began to fall apart.

There was my marriage, or rather the end of it. Connor informed me he had realized he was in fact gay, not bisexual. He had grown up in a five-thousand-person Illinois town and had primarily dated women throughout his twenties, and the homophobic context of his upbringing made it difficult for him to imagine being gay. Once his best friend came out as gay, he realized he hadn't been honest with himself, or with me. There was nothing left for us to do but separate. I always try to remind people homophobia is bad not just because it hurts people who identify as queer, it also hurts those around people who identify as queer. We as a society don't want to pressure people to live a lie, because all lies are liabilities yet to be paid.

We separated with love and support for each other—over the next few years he never let me go without health insurance as my physical problems worsened or left me to face challenges in isolation, despite moving into his own space up in San Francisco. I was now alone in Mountain View. Until this point in my life I had coped with unhappiness or adversity by burying myself in my work, but now, with no work but slide decks that I might as well have just put in the shredder myself, it wasn't clear what my work really was.

It was under these conditions that I made one of the most ill-conceived decisions of my career. Instead of staying with Google+ and begging to be used as a backup product manager under another PM and contributing to their project, or just remaining in limbo, I went and found myself a new team and informed my manager that I wanted to transfer. My manager looked at me very kindly. Maybe a month earlier, I had started to experience symptoms of my worsening health—bouts of dizziness, which one morning caused me to be late to a meeting with the Google+ leadership. Vic Gundotra, the VP of Google+, inspired strong feelings at Google for a variety of reasons, but I have always been a fan of his because he was the only person who actively encouraged me to seek medical help during my two-year decline. I was an invincible twenty-seven-year-old, but he was my boss's boss, and he gave his admin the task to nag me until I saw a doctor.

Now my manager asked me, "Are you sure you want to do this? We know you here, we know what you're capable of. We can look out for you." This is what a good manager says, even though it's a net negative for his team, and I absolutely should have stayed on Google+. I told him I thought that if I took a month of medical leave, I would be back in good shape. He slowly shook his head; it was all he could do. Unlike Lucas, he was my *actual* manager, not just a mentor, and as a result he was limited as to what he could ask me about my health issues. He knew the truth of the matter, though.

One of the challenges of going on medical leave is that you occupy a head count on your team, which leaves it short staffed. I would be the second product manager on my new Google Knowledge Graph team, and by going on leave I would be taking half of their product manager capacity off the table, a massive reduction. Taking up head count on Google+ would not have been an issue with my manager or the Google+ team as a whole. I knew firsthand that we had a cubicle full of product managers with no work to do—if anything, my going on leave would have been an advantage for the org because it would have protected the head count from being cut in Google's quarterly efficiency housekeeping.

But my manager couldn't explain that to me without crossing legal lines that could put him or Google at risk of a labor lawsuit. And for my part, I felt too vulnerable to be willing to accept help. Too much of my identity was wrapped up in my job, and being good at my job. Too much of what I valued in myself came from what I produced at work. On Google+ I was only producing vaporware, so I was anxious to make the move. I thanked my manager for his thoughtfulness, but maintained that I wanted to go. The new Knowledge Graph team was up in San Francisco and had concrete work to be done. Joining them would shorten my commute to my new home in Oakland.

What I naively didn't register was that for the first few months you're working on a project with a new team, you're not only having to come up to speed on a new space, you're also setting the receiving team's impression of you. Are you a hard worker? Do you pull your own weight? Coming into a team when you're frail enough that you have to go on medical leave for a month beforehand does not exactly set you up for success.

This single choice was the beginning of the end of my time at Google.

My neuropathy was making commuting ever more taxing—I still to this day struggle with the vibration of cars or planes; the added physical noise makes it harder for me to ignore the noise already present on my nerves. I wanted to minimize my commute to Oakland so much that I ignored or blocked out enough of the red flags as I joined the new team. The first and most glaring was that I was not co-located with my engineers. My desk was in San Francisco, and while members of the broader Knowledge Graph team sat around me, the actual team I supported was in China. I was in the United States because my job was to liaise with and support projects across Google that wanted to use media metadata (music, movies, books, and so on) and connect their needs with the engineering team in China.

This left me in an awkward onboarding position in some ways similar to what I would experience years later when I joined Facebook. The Knowledge Graph software was changing so fast that none of the documentation was up to date in even basic ways. I could ask those around me for help, but I was not their product manager. Every question they answered would not eventually benefit them once I was up to speed—it was just a pure favor to me—and I knew I would soon be annoying them if I asked for help too often. The team I actually supported in China didn't come online until five or six in the evening California time.

At Google, officially your value to the company is assessed through peer and manager reviews. But on the most basic level, one implicit way your value is assessed is whether or not someone is even willing to write you a review in the first place—each person back then was only obligated to write seven or nine peer reviews at the most, unless they had a more senior role. For obvious reasons, I was concerned. My role on any of the projects I liaised with was a small portion of the overall effort. I warned my manager early on that I was being put in a position where the people I spent the most time with would likely not write me a review even if I did a good job because I was being spread across too many projects. Despite the impact of the neuropathy meds, I felt especially run-down. I didn't know why.

It wasn't as if I had much to look forward to at home, either. By 2013 I had moved into the Oakland house, though it was not fully functional and still lacked heat. I now know my neuropathy is exacerbated by cold, but back then I hadn't really put two and two together yet. All I knew was I came home to an empty cold house each day and would curl up in bed with my electric mattress pad to keep me warm. I only saw my co-owner friends when they came over from the sister house next door to do laundry.

The renovations had been grueling, over budget, and behind schedule. I knew my salary at Google was what kept our mortgage and contractor bills paid, and the stress of that responsibility made every tough day at work even harder. We were close to being finished, but not quite enough to be out of hot water with our bank. The only way we had been able to get a loan on our condemned property was by working through a program with the Federal Housing Administration, and the consequence for not being officially habitable in the near future would be to lose the property.

Every day I came home to a house that only had one floor fully finished, and the small yard between our two houses became a chasm that we seemed to cross only when we were sufficiently frustrated with each other.

———————

And just like that, the situation got much worse. I was tipped off by a friend that a Performance Improvement Plan (PIP) was coming my way soon. I decided to go back on medical leave, this time for three months, in the hope that I could recover enough to pull myself together sufficiently to weather a PIP and dodge being terminated.

I finally had to confront just how relentlessly I was struggling physically. My therapist wasn't one to give me much direction; she was the type who wanted me to find the answer on my own. But one day she lowered her clipboard, looked me in the eye, and said, "You have a giant hang-up about needing a wheelchair." I looked at her like she'd just invented the lightbulb. This had been right in front of me, and now I could see it clearly. I went online that night and bought a wheelchair from a start-up founded by former engineers of performance bicycles. The day it arrived Connor and my brother, Peter, who was studying nearby at Berkeley, took me to IKEA so I could roll around on the slick polished concrete floors. I wheeled in and out of every display living room and dining room, finally feeling free after having been confined by my body for months. Connor and Peter hadn't seen me truly happy very often that year, and they clearly relished giving me a little joy.

It was within this context that I made one of the worst mistakes of my life. It was now very apparent my manager wanted me gone. I was told by a friend that he allegedly thought I was faking the need for a wheelchair. He had patiently waited out my medical leave—which I should have stayed on until either I got fully better or until the time available to me ran out—and he informed me it was time for my PIP. I tried for a month to work through it, but it was becoming apparent that it wasn't a matter of if I would get canned, merely when. One day, perhaps soon, I was going to get a mysterious invite on my calendar announcing the inevitable.

I accepted an offer to take a severance agreement and left the company. I didn't know what was about to take place, and I didn't know how valuable it would have been for me to have a guaranteed job when my medical ordeal was over. What I did know was that I needed the money. We had squeaked across the finish line of our renovation requirements with the bank just as the money was running out. If I was fired, I'd get nothing, so I signed the papers and left the place that had represented so much of my life, so much of my identity, for so long. Google was the first place I ever felt I really belonged, and now, for the first time in over seven years, I was no longer a Googler.

Rather than be alone in the Oakland house, I rented a small apartment nearby in Emeryville. I justified the rent by telling myself it was going to be my office. It would be a place to go during the day so I didn't feel like I was sinking into a hole in Oakland. That was until the day I got the keys and stood in the partially lit and empty space. The feeling of release was instantaneous. The apartment wasn't anything to write home about; it had a beige carpet and white walls, impersonal as impersonal could get. Not exactly a home like the one we'd been trying to build in Oakland, but that one-bedroom unit felt like hope. It was a room of my own. Within a couple of days I had acquired a mattress and a secondhand couch.

I had my own place now, but I had no job. I co-owned a house that was costing money, and my co-owners and I were increasingly at odds. I felt lost in my own tiny space, alone. And being lost was how I passed the next two months, with occasional breaks to stay with Connor in San Francisco. Until one morning, when I woke up at Connor's and discovered something extremely disconcerting: one of my legs had turned blue.

CHAPTER 8

Navigating with Numbers

Dumb AI is a bigger risk than strong AI.

— VentureBeat

The first thing I did was call my parents. Without thinking, reflexively, I reached out to ask my father's opinion. Having a doctor in the family is the ultimate amenity when you're staring at a blue limb attached to your body. I was still in the phase of shock where your mind doesn't let you imagine that this could be serious.

On the phone, my parents were sensible; I was not. They urged me to go to the emergency room. I suggested that maybe I should go to the massage I had scheduled for that afternoon, and then go to the ER, not realizing how avoidant I was being. "Unquestionably do *not* get a massage," my mother said. "You could dislodge what might be a clot in your leg and give yourself a fatal embolism." She was adamant that I needed to get myself to a hospital. She was right. I had a deep vein thrombosis clot. Although this can be very serious, for most people, it's not. My mother had had one, and after a week of bed rest, she was fine. That was everyone's initial hope for me, and this provided the first of seemingly endless times I heard, "Don't worry, you'll be home in three days."

Instead I embarked on a fifteen-month pause in my career involving multiple operations and close to two months' worth of hospital stays, complex and shifting diagnoses, a medication that accidentally killed off my platelets and brought with it the complications you can imagine that might entail, and a long period in which I could only get around with a walker or a cane. It turned out that I had not been eating what I thought was a gluten-free diet in Google's San Francisco cafe, and my muscles had lost the ability to repair themselves. I learned the hard way that the statement "I make a hundred and eighty dishes a week, I can't be expected to get the allergy warnings right" has real human consequences. If that wasn't enough, I was so weak at one point that I fell and hit my head in my bathroom, sustaining short-term brain damage that meant I couldn't even talk for a while. When I thought I had escaped the worst of it, I discovered the odd gait I had when I walked was because I was paralyzed from my knees down, and had just learned to compensate over the years.

Blowing the whistle on Facebook was not the hardest thing I've ever done. People give varied reasons for why they can't follow their hearts. They're afraid of losing their jobs. They're afraid of losing what wealth they have. They're afraid of being ostracized and alone. None of these scared me away from blowing the whistle on Facebook because I had already experienced them all, and at the same time. This era, and my recovery and return to living a vibrant and fulfilled life, is a story about how we stand up when we find ourselves brought low. I hope one day I can share the story of how I claimed agency in my own life, because learning I could get back up when I fell down has been one of the most liberating lessons I have ever learned. But in the spirit of keeping this book to a manageable length, I'll save those details for another day. Suffice it to say, I hope you never find yourself where I did in 2014 after countless medical interventions. The important part is that *I recovered*. It took more work than I could have imagined at the beginning, spread over many years, but it happened. And my parting nugget of advice, if you ever need to relearn to walk, is that a tango instructor can be more effective than a traditional physical therapist.

Beyond the dance instructor I waddled around the dance floor with, the other person who unexpectedly changed the trajectory of my recovery while I was literally getting back on my feet was an administrative assistant I hired who became a close friend. I met Jonah in March 2015 when he was living with my brother, who had moved to San Francisco from Berkeley and enrolled in a Developer Bootcamp to become a programmer. Peter and Jonah were living in a Silicon Valley hacker house called "Hacklantis" with about a dozen male roommates who were all trying to make it in tech.

It was little more than a converted industrial garage full of bunk beds and desks, with a bathroom and a kitchen tacked on. Such living arrangements were not unheard of in Silicon Valley. People from around the world flock to San Francisco hoping to be the next Sergey Brin, Mark Zuckerberg, or Evan Spiegel. They resign themselves to a lifestyle that would be considered ridiculous anywhere else in the United States.

I hired Jonah to do some odd jobs for me that I still wasn't physically able to handle, and we became friends. When he moved out to Berkeley and into living arrangements that stimulated my protective big-sister Spidey sense, I offered him a trade. In exchange for twenty hours of assistant work a month, he could use my apartment as an office while I was working in San Francisco. My place was clean and safe, and I wasn't going to be there. And frankly, I really needed the help.

When I came home at the end of the day, we'd often walk to the Emeryville marina along a path that clung to the shoreline of the San Francisco Bay. It was a true act of patience in the beginning; the normally thirty-minute-each-way walk would take me forty-five minutes or even an hour, and I needed long rests on the benches that lined the path. Naturally, we talked at length, about anything and everything. Jonah was smart, empathetic, and scrappy. He was a dedicated gym-goer, knowledgeable about fitness, and a great coach and cheerleader. He had graduated from college in three years and then gone to Taiwan, where he had worked for an entrepreneur at his start-up. I was grateful for his friendship and willingness to help me rebuild my life and came to think of him as sort of another little brother. My connection to him would prove to be a pivotal motivator for what I did at Facebook, but more immediately, Jonah was critical in my rehab and why I felt better about returning to the tech workforce, this time at Yelp.

In the spring of 2015 I had been out of the job market for fifteen months. I was running out of money from my Google severance and had wiped out my savings in the bottomless money-pit that was the house renovation. I was operating under unnecessarily difficult constraints as I heeded a commandment that prevented me from touching my retirement fund under any circumstances — the chorus of finance influencers who constantly droned, "That's for your *future!*" When the job offer came in from Yelp, I was grateful that it offered a paycheck.

I returned to work in late April 2015, only two weeks after I had stopped walking with a cane. I remember on my first day of work I had to make the half-mile walk from the temporary Transbay Terminal to Yelp's offices. I had struggled to complete the distance and arrived sweaty despite the chill air. One of my greatest regrets is that I forced myself back to work rather than wait a few months more and return far stronger. I offer you a hot take: sometimes the best way to invest your retirement funds (if you have them) to ensure a stable future for yourself is by caring for your present self. In my mind, I had to get a job, *now,* and I was blind to the long-term cost.

A few months before, to prepare for returning to work, I began picking up hobby projects that I could use as work samples for future employers. This also was the ideal remedy for my growing boredom — I had recovered enough mentally that sitting on the couch, anchored by my still healing body, was becoming a burden. I missed working on Search Quality and set out to identify good gluten-free restaurants by combing through Yelp.com's review data using a scraper and data processing pipeline I'd written. Yelp had been around since 2004, the same year Facebook was founded, and had amassed a treasure trove of information about restaurants and other small businesses. I was particularly interested in the details people disclosed about themselves in their reviews. A reviewer might talk about their own gluten-free preference, or the celiac diagnosis of their spouse or child. These shreds of information colored the credibility of the reviewers. If you want a sufficiently rigorous gluten-free review, I'd take one written by the mother of a gluten-free child over someone who happened to be dining out with their gluten-free friend any day. Mama Bears ask all the questions, and they'll let you know if they don't like the answers.

Going from Google to Yelp wasn't an obvious transition, particularly with a background like mine, but the way my manager on the Knowledge Graph had pushed me out had left me deeply skeptical of my own abilities. I was just grateful when I got the job offer. It was a massive pay cut from Google, but they warmly welcomed me and were enthusiastic that I wanted to work on Search again. They were giant data nerds and loved my gluten-free-restaurant search engine.

That spring into summer felt like a gift. I was assigned two teams: the Suggest team, which focused on results that autopopulate the search box, and the Data Mining team, which used the mountain of data Yelp collects to build useful product features. When I arrived, there was plenty of low-hanging fruit and an opportunity to do good work. It felt nice to have a team that wanted me to be there, and I was thrilled to be back working with search data. Here again was a digital Russian novel showcasing the human condition. The majority of content on Yelp is for restaurants, but every corner of a person's life from cradle to grave, from grace to sin, is present within Yelp's search query data in far more permutations than you could ever imagine. We didn't have to try hard to find little haiku-like search sessions that sent our imaginations racing, they'd just show up over and over again as we debugged our algorithms. Who was the man (and we presumed it was a man) in Chicago who was reading reviews both for day care centers and escorts? How bad was the shedding problem that person's dog had, given the hundreds of dog groomer reviews they had read through? What kind of person (or couple?) looks primarily at reviews for steak houses *and* vegan restaurants? It reminded me of a statement made by a presenter at a Computer–Human Interaction conference I once attended: "We are all social scientists, studying the human condition through the medium of the internet."

I had left Google right as it was entering a reinvention period. Machine learning, a subset of the broader field of artificial intelligence, had been an academic field for years, but in the mid-2010s, the machine learning algorithms for helping computers tease apart complex patterns from large datasets were finally fast and cheap enough that Google was considering rearchitecting the Google.com search engine to take advantage of the technology.

Starting right as I went out the door in January 2014 and into 2015, I read about how Google was running their employees through boot camps to teach them how to use these new statistical tools and how to solve problems using machine learning. I read these pieces with a great deal of regret. Lucas and I had been part of the small vanguard of product managers who worked extensively with data in the mid-2000s, and it had perfectly set up him and those on his team to take advantage of this new era of technology. I, on the other hand, was stranded out in the wild. *Had I missed the boat?* I wondered.

Then one day at Yelp one of the engineers on the Data Mining team mentioned he wanted to build Yelp's first machine-learning feature using a technique called computer vision. He wanted to classify the large number of photos on Yelp's restaurants into food, interiors, outside spaces, and of course menus. I jumped at the opportunity—I knew I could help him iterate on his algorithm and turn it into an effective user experience.

Of all the things I've worked on, the menu filter on Yelp is the one that most often makes people say, "Ooooh! I use that!" It was surprisingly hard to get approval to build, I think partially because of misunderstandings about how people actually used Yelp's products.

Yelp was where I came to appreciate the fundamental topology that drives all social networks—which is to say, regardless of whether you're talking about positive patterns or patterns of bad actors, there are repeating patterns of network behavior. One simple example of this is the 90-9-1 rule of Participation Inequality in social media and online communities. In most online communities, 90 percent of users are lurkers. They consume content but never contribute; if you want to be mean, you could say they leach off the participation of others. Another 9 percent will engage in lightweight actions like a "like" or a comment. And 1 percent of users create all the content the other parties consume. That distribution is a "long tail" pattern. In a given month, that vast majority of users will post not at all or once. A small minority will post a few times. Among the posters, maybe the 1 percent most active will post one or more times a day.

When it came to photos for businesses on Yelp, the businesses followed a similar pattern. I remember asking the product manager why the business page didn't incorporate more photos into its design, and he replied that "There's no need—ninety-nine percent of businesses have zero photos." That was almost certainly true—of the businesses Yelp knew about, most were not restaurants, the original category Yelp was founded to review, and people were much more inclined to photograph their sandwich from lunch than the waiting room of their podiatrist, no matter how stylishly decorated the waiting room was.

The product manager for the business page was thinking from the perspective of Yelp's data, not of what users were actually viewing and using. Most businesses on Yelp aren't viewed in any given month, while a small number are viewed thousands of times a day. Even within a single category, like Sushi Restaurants, some restaurants are viewed many times a day while others might get viewed every other week.

What drove these differences in user consumption? In the case of people who came to Yelp to find out about a specific restaurant, some restaurants were more famous, or at least known and preferred, than others. You wouldn't expect all restaurants to be equally sought after. But for people who came to Yelp wanting to be guided by Yelp's search engine to what they should have for dinner, the algorithm was suddenly left in the driver's seat and responsible for much of the skew. The search engine at Yelp rewarded businesses that had more reviews and photos with higher search rankings, which created a cycle where the rich got richer. More people found the businesses that had more photos and reviews, and this in turn meant that more people could take photos and write reviews. In this way, Yelp's choice of algorithms directly affected businesses and, in a way, limited information and (potentially) good options for Yelp consumers. Cuisine that is more unique — say, Thai — is more likely to generate a review; and food that is more visually unique is more likely to get a photo. "Notable" isn't inherently synonymous with "good," but the biases in Yelp's algorithms directly influence what Yelp users tend to see. In this way, Yelp was not uniquely good or bad — every single algorithmic system has some form of bias, intentional or not. But because Yelp's algorithms are not published and Yelp declines to share appropriate performance data on how those algorithms operate, Yelp is driving the success or failure of small businesses around the world without accountability or oversight from the public.

I thought we should be driving the design of the business page based on how users were consuming it, not on what database was sitting on a hard drive in a data center. When it came to photos, if you lined up every restaurant business view (think one load of the web browser on a business) left to right, based on how many photos the business that was viewed had, you would find the fiftieth percentile (median) business view was of a business that had fifty-five photos. I understood why the business page product manager was afraid to use photos more aggressively, but at least for restaurants, we were leaving a giant opportunity on the ground.

My time at Yelp helped me appreciate a number of shibboleths I had or would encounter throughout Silicon Valley, and most definitely at Facebook. It's easy to criticize Facebook management for its sometimes blind attachment to navigating almost solely with numbers (and I'll certainly get to the consequences of that management style), but there was a reason why Facebook steadfastly clung to the idea that the only way to manage their operations was to set objective quantitative goalposts and let people run free as long as they moved the ball farther down the field. On the most basic level, according to the metrics of market share and revenue, it worked. Meanwhile many other tech companies that have cultures unmoored from metrics descend into politics and favoritism, and ultimately failure. Facebook wanted to free their employees to move forward against those internal ossifying forces. They knew young people drive the future of social media, and a culture driven by metrics treated a fresh graduate the same way as someone with ten years of experience. Did you make the metrics go up? Good. You win.

Yelp was in a risk-averse phase, having undergone a major redesign a few years before that had been developed with the aim of unlocking user growth that never materialized. By the time I showed up it was explained to me with a hand-wavy level of detail that they had swung hard to hit a home run and failed. Now they only seemed to feel confident about growth hacking and reoptimizing how Google indexed their business pages and search results to influence how often Google presented links to Yelp on Google web search. Growth hacking was a sexy way of describing looking for small, easily measurable tweaks to the user interface to drive core metrics up. One of the reasons tech companies are so obsessed with this style of incremental growth is that it allows product teams to focus on hitting easily measurable singles and doubles; you can legitimately improve the profitability of your product by 10 or 20 percent a year just by removing friction and cleaning up poorly executed features. But you're unlikely to truly innovate and unlock new user bases or use cases this way. The Yelp Data Mining team was supposed to be a research team unlocking new capabilities in the product, yet it felt like pulling teeth to get permission to do even minimal experimentation.

I started to wonder how Yelp was able to hire so many high-quality people. Strong technologists are generally attracted to fast-growing environments where you can create (and therefore claim) value and success more easily. Yelp was not one of those places. The company while I was there was not actually profitable according to Generally Accepted Accounting Principles, a fact they tried to hide from their employees by saying things in all-hands meetings like, "We added XX million dollars more to our bank account last quarter" while glossing over the fact that they had issued enough employee stock — something they were required to count as an expense in their quarterly reports — that it wiped out the profit. Yet they continued to hire terrific new talent.

Soon after I joined, I figured out how they did it. Back in 2015, Yelp was one of the only tech companies that made a dataset of its reviews freely available to universities for academic research. Every year they would release hundreds of thousands of reviews coming from a handful of cities (none of which were their top competitive markets), and the dataset would be used in computer science courses around the world to teach natural language processing and other techniques. When it came time to pick internships, students would think back on how much they enjoyed working on the Yelp dataset and would apply to intern there. This created a steady stream of talent for Yelp, far beyond what the fundamentals of the business would suggest.

While most big tech companies invested in their intern programs, Yelp went above and beyond to make sure interns had a worthwhile summer with events like baseball games and paintball outings multiple times a week. Before the summer internships ended, Yelp asked participants to write themselves a letter describing how much they had enjoyed working there and how much fun they had with their new friends. Then Yelp recruiters would drop those letters in the mail just a few months later in the fall, timed to arrive just as the graduating students were considering their full-time job offers. The best person to pitch yourself is yourself.

I think most companies look at demands for transparency with resentment — it feels like a slap on the wrist, that the public doesn't trust them to do the right thing on their own. How would these companies be different if transparency were viewed as a strategic advantage? Yelp built a pipeline of people already familiar with their business by being willing to open the curtains just a little bit. What would it take for other social media companies to get the same advantage?

In the wake of my whistleblowing, I've invested substantial time to think about how we can drive systemic change of social media platforms. To support that work, I'm founding a nonprofit called Beyond the Screen devoted to the dream of expanding by one million people the pool of those who rigorously understand how social platforms work. One of the major projects we're going to work on over the next few years is creating simulations of social networks so that we can provide a lab bench for researchers, the next generation of social media technologists, and everyday concerned citizens to imagine what's possible online. Imagine if Facebook or Pinterest could release configuration files and tuning parameters that would allow a simulator to make a facsimile of the virtual worlds that they operate every day on their platforms and the challenges they wrestle with. We could generate industrial-scale datasets that would let students really see how complicated their future jobs would be — and with simulated data, nobody's privacy is violated. How much would it help social media companies to have a stream of students interested in their companies who were already familiar with how their systems were structured, or the resulting data they threw off? Freshly minted algorithmic engineers or data scientists take years to fully ramp up — how would social media companies benefit if they could hit the ground running on day one?

What allows society to provide effective oversight for an industry is not just standing up a federal Department of Industry X in the government. We need to have enough people, enough different kinds of people, spread throughout society examining how the industry operates, all with their own different skills, perspectives, and interests. A world in which Facebook had to publish their recommender system architecture and machine learning model tuning parameters to generate a faux Facebook is a world in which academics and civil-society groups, alongside YouTube Influencers and the like, could discuss how the systems operate and how to creatively fix the biases that naturally creep into any algorithmic product. We don't have to let Facebook struggle to fix their problems by themselves—we can do this collaboratively. Today, because no social media company grants anything close to this level of transparency, you cannot take even a single class on how to architect a social network or weigh the tradeoffs that come with each decision you must make along the road of development. The only people who understand these systems were trained within the companies. Unlike every similarly influential industry, Facebook cannot rely on an army of academics and graduate students constantly tinkering and developing solutions to the problems Facebook faces. By allowing social media companies to keep the curtains closed out of short-sighted fears and the real challenges/costs of being the first mover on transparency, *we're choosing to let them struggle and fall short, isolated and alone.*

The year I spent at Yelp ended abruptly in the summer of 2016, when one of my closest friends, Simon, called to tell me about a job he thought I should apply for. He was working at a relatively new, at least compared to Yelp, start-up called Pinterest. Simon had ended up at Pinterest because he had been childhood friends with Ben Silberman, the founder and CEO, back when they were both debaters in Iowa. Ironically, I had met Simon over lunch one day back in 2008, when Ben invited me to eat with the two of them at the Googleplex while Simon was visiting from Harvard Business School. We were three Iowans, thousands of miles from where we were born.

Given the much-higher-paying job dangling in front of me and the promise of working at a growing start-up, I jumped at the offer. I left for Pinterest, and for my real education in machine learning and the biases that inadvertently slip into algorithmic systems along with the consequences of those biases. The work I did at Pinterest gave me the experience and expertise I needed to understand what was wrong with Facebook's systems and the confidence to blow the whistle.

The first six months at Pinterest were a crash course in a very different way of thinking. The practice and study of extracting information from a database is called information retrieval. To grossly simplify a complex field, information retrieval systems are divided into search engines and information discovery systems (also known as recommendation engines). Search engines are for when you're expecting a specific answer or answers to a question. Think of searching for [what is the airspeed velocity of an unladen swallow?]. You might expect a response like "Did you mean European or African Swallow?" or you might assume you'd get a number in units of miles or kilometers per hour. You know what you're looking for, and you'll be disappointed if the search engine responds with something that feels random.

Discovery engines/recommendation systems start from the presumption that you don't know exactly what you're looking for, you just want some suggestions. The book interface I built for Google was in some ways a form of discovery engine for information within a single book—it was meant to help you understand where to focus within a book, even if you didn't know where to start. In the case of Pinterest, the platform offered a simple user promise: If you tell us what you like by pinning (that is, adding images) to boards of things you're interested in, we'll recommend more things you might like. The "query" used to retrieve that information is you and your interests.

Search engines have been around since the 1950s and '60s, beginning with the move to digitize library holdings and corporate document catalogs. Recommenders radically changed the field of information retrieval and our overall information environment. Now a computer would tell you what it thought was relevant to you. If Google was built on decades of research around how to think about search, Pinterest was something fundamentally different. Its first version was launched in 2011, and was built on the thinnest shell of existing research and knowledge.

Search at Google had an "objective" way to evaluate whether one iteration of its software was better or not: at least five hundred queries would be identified that yielded different results, and at least three but up to seven to nine raters would be shown each pair of search results. Each of those raters would have a manual tens of pages long that in minute detail described what made search results "better." With every tiny incremental change to Google's algorithms, the opinions and perspectives of human beings directed the search engine forward.

Recommenders, because they were personalized, were inherently *subjective*. We couldn't evaluate our recommendation engine at Pinterest the way Google did because what a recommender presents to you and to me will be fundamentally different—we each have access to different information, and the system will prioritize it based on our individual interests and preferences. Lacking the ability to ask humans "Are we doing better?," we were forced to find other ways to describe whether one version of the Pinterest recommendation engine was better or worse than another.

This started out simply: If one version led to more repinning (adding a pin to a board), it was better. But subtle differences in the metrics we selected could end up detecting different kinds of changes. Take for example if we counted and prioritized how many *total repins* happened. A change that made heavy users of Pinterest happier but the average user slightly less happy might increase total usage because heavy users might repin twenty or a hundred times as many pins. If we counted and prioritized how many *people* repinned pins, a change that made the average user happier but heavy users less happy would likely increase the *number of people* who repinned while potentially marginalizing our best customers.

By the time I was there, we were tracking a hundred or a hundred and fifty different metrics which we could further slice and dice until I could answer questions as specific as "How did 65+-year-old Spanish-speaking men in France respond to this change?" We didn't know what the "right set" of metrics was, we just kept adding more and more to detect additional edge cases as we found them.

I can speak critically of how Facebook allowed itself to be led into the wilderness with metrics while still being empathetic about why they did it because I've lived through and understand the real need to regard metrics as a North Star. Without at least that thread to lead you forward, you just get lost. The real question is who gets to imagine and check for edge cases. We will never determine a perfect set of evaluation metrics—one that will never lead us astray; we can only continually inspect how our electronic shepherds lead us forward and make sure it's a path we desire.

In 2016, when I joined Pinterest, the top metric that represented success for either the Home Feed or Related Pins was the "total repin rate." It was aimed at the core of what Pinterest was. The name Pinterest is the combination of pin board and interest. We had countless ways we could analyze those pin patterns: for example, which *type* of pin had been pinned (a single image, a video, a collage, and so forth), or what type of *user* had viewed the pin (age, language, country, how new they were to the platform, and so on). These are slivers of examples of the almost limitless ways Pinterest could break down the metrics.

Even the question of *whom* to prioritize was up in the air—do we optimize the system to maximize your individual happiness? Or do we recognize that your participation in the system will directly influence the happiness of others? We might then care less about your individual happiness or well-being in exchange for the system's success. This conflict between optimizing for the individual and optimizing for "the network" was core to the Meaningful Social Interactions controversy that arose with my Facebook disclosures. Facebook had rolled out a major press campaign in 2018 when they switched from optimizing for keeping users on the system for as long as possible (thus viewing and clicking on as many ads as possible) to maximizing what they termed "Meaningful Social Interactions." Facebook was explicit in their press release—they cared about their users, which is why they were now rewarding content that got comments, likes, and reshares—they didn't want you to mindlessly scroll. Facebook's documents revealed that their real motivations were different.

Many were outraged when they found out that Facebook's move to "Meaningful Social Interactions" was not done to make people happier—people reported six months later that their feeds were *less* meaningful—but because generating more interactions stimulated the creators who *received* those interactions to *make more content*. Facebook had switched from optimizing for the happiness of users (how much time they spent) to using them as a means to an end (to generate likes, comments, and reshares that made creators create more).

If you had asked the average person on the street, "Has Pinterest changed in the last ten years?" I think most people would say no. Just as if you asked the average person, "Has Facebook changed?" they would also say no. A feed of pins or a feed of posts can't really look that different, except when you add a new format like images or video into the mix. But the reality was the Pinterest of 2016 was very different from the Pinterest of 2011. Pinterest had begun as a website that was almost like an email inbox for people whose taste you liked. You followed them, and when they found something they liked, it showed up in your feed. At least initially, people directed your attention, not computers. Pinterest also provided a search engine for finding new things. Repinning was important, because each time you repinned, your content would flow downstream to all your followers.

With time, the motivation for stimulating repinning evolved as the product grew more complex. The next major change came in the form of Topics, subjects that users followed in lieu of finding fifty-plus people to follow to fill out their Home Feed. This radically amplified some content— now a repin wouldn't just go out to your followers, it might also go out to hundreds of thousands or millions of people if you repinned the pin to a board with a name that matched a known topic.

Then Pinterest made a quantum leap forward: pure recommendations. You didn't know why you received that pin, Pinterest sent it to you just because the *algorithm thought you would like it*. It was a black box; even Pinterest's engineers likely couldn't tell you exactly why you received it. Now each act of repinning taught us a little bit about your preferences, the nature of the pin itself, and who else might also like it. This invisible evolution is part of why we need transparency for our algorithmic environment. Even if a product or experience begins in an innocuous way, it can shift and evolve over time into a very different system. When many people originally joined Pinterest, they put their feeds in the hands of people they trusted. Ten years later, most of those people had feeds dominated by the whims of an algorithm. How many people realized it?

Our laws are similarly blind to the changing online world. Today, as I write, the United States Supreme Court has taken up a potentially transformative case in *Gonzalez vs. Google*. In November 2016 a woman in Paris, Nohemi Gonzalez, was killed by an Islamic State attack. Her family sued

Google, which owns YouTube, claiming that Google's algorithms recommended Islamic State propaganda to its YouTube users who would have never sought out the content on their own. Google knew it had proterrorism content on its platforms and did not act.

At the heart of the case is Section 230, a statute central to the birth of the internet. Passed in 1996 as part of a larger Communications Decency Act, Section 230 provides immunity to online companies like Google, Facebook, and Twitter for content they make available on their platforms. Google's position is that it does not intend to publish terrorism content and it is impossible to entirely prevent the distribution of such topics. If they did not have immunity, they would have to stop hosting wide swaths of user-generated content because the incremental risk of a lawsuit would outweigh any business value gained by hosting the content.

The fact that the Supreme Court is hearing the case at all is a symptom of how Section 230 is struggling with a world radically different from the one in which it was written. When Section 230 was written, there were no recommender systems in the wild. Only a few academic research projects existed for filtering email. The first commercial recommendation system appeared on Amazon three years later, in 1999, when it began suggesting additional items you might like to buy based on past purchases. Facebook launched their algorithmic feed in 2011, and it took Twitter twenty years after Section 230 was passed to launch their first algorithmic feed.

Why does this matter? When Pinterest, like Facebook, first introduced its feed to the public, Pinterest was clearly tightly aligned with the existing interpretation of Section 230. Facebook gave people an ability to opine on the state of the world or their lives—on anything, virtually in any manner they chose—and those feeds spread that content out to the people they *knew,* who back then were largely their *actual friends.* It was like email in that it was a very simple delivery channel. What came in flowed right back out along relationship lines that users controlled. If you received harmful content, it was because you *chose* to be connected to the person who sent it to you. Facebook's differentiating trick was that it enabled many-to-many conversations more effectively than email did.

Where the world of Section 230 changed was when Facebook started making decisions about which pieces of content they promoted. You could say something outrageous and have it go viral, but if the same person posted a fact-based rebuttal explaining why that post was nonsense, that correction would almost never be seen by as many people. Facebook might even choose to put content in your feed you never asked for—like a post from a group you were invited to, but never joined. *Gonzalez vs. Google* asked the question, "If Facebook can control the algorithms it uses to route content, to influence what you do or don't get to see, should they be responsible for those choices?" People had signed up for a product that was about connecting with their friends and family, and it shifted under their feet until for many, it became something else.

Back in 2016, I was living the context that would lay the groundwork for *Gonzalez.* We were part of a vanguard of practitioners learning about the limitations of recommender systems. Unless we actively went looking for an unintentional harm, developed a metric to measure it, and then consistently used that metric in assessing experiments as a protective guardrail, the harm was invisible. Much of the work today around introducing regulatory frameworks grounded in mandatory risk assessments, like the EU's Digital Services Act, is intended both to require a process of reflection that might otherwise be overlooked, and to allow the public to understand the blind spots of tech companies and correct them.

It was in this context in the late summer and early fall of 2016, as the energy, vitriol, and toxicity began to swirl around the presidential election, that I began to take a hard look at Facebook from afar. This was almost entirely because of Jonah.

It had been at least a year since I started renting him thirty square feet of my living room for an office, and often after I would get home from work at Pinterest, Jonah would still be working in my apartment on the application he was writing. We had kept taking our walks, and it was during these lazy strolls that I started to notice something changing within him. He had been an enthusiastic Bernie Sanders supporter that spring, and when Bernie lost to Hillary Clinton, he had turned to online spaces to commiserate about how the primary had somehow been stolen. Starting from that nugget of grievance, he began going deeper and darker, as algorithms pushed and pulled on his anger. By the fall we were regularly exchanging long emails regarding some of the conspiracy theories he had begun to share with me. I felt Jonah was slipping into a place I couldn't understand and I couldn't reach him with reason. It was a microcosm of the problems that social media were inflicting on us all.

I felt powerless. Jonah had been a loyal friend throughout my recovery, and to this day I still feel I owe a portion of that recovery to his dedication to getting me better. Watching our realities drift farther apart sensitized me to misinformation I saw distributed when I would log into Facebook. Just glancing at certain features made it obvious that not enough people were trying to ensure that the core of the system would not be compromised to disseminate false narratives and propaganda.

One of my main projects at Pinterest focused on inherent biases in our system regarding older content. Unlike Facebook or Instagram, Pinterest makes no user promises about the freshness of the content. A data scientist had discovered that the average age of pins being shown was steadily increasing, which had the side effect of discouraging new content creators from adding content to the system; just as Facebook discovered "likes" and reshares could stimulate content creation, distribution and repinning stimulated content creation on Pinterest. By optimizing for repinning, we were inadvertently biasing toward older content, because older content had accumulated more helpful information or "signal" on what the pin was about or who might like it. An older pin was a less risky pin to the blindly optimizing hill-climber that was our recommender system. A simple algorithm on its own couldn't know it was making short-sighted decisions that might deprive it of new content in the future.

As I looked at Facebook it became obvious to me that the folks there were not thinking about what biases their algorithms might be introducing or the consequences of those biases. One feature in particular caught my attention. Whenever you clicked on a Facebook link in a post — or clicked "Like," left a comment, clicked out on a link and came back — a little carousel (a horizontal stream of posts) would appear under the post with thumbnail images and link titles of other related news stories you might be interested in. No matter what I clicked on or "Liked," even the most ordinary stories, the carousel served me stories with absurd headlines like "Pope endorses Donald Trump for President." I knew clickbait when I saw it, and it looked like no one had noticed that this carousel had become a megaphone for the most sensationalistic content.

When the 2016 election was over, it seemed like my friendship with Jonah had reached a critical point. Maybe a week after the election, Jonah and I had one of our last extended email threads. I was trying to debunk his statement that "George Soros funds violent revolutionaries." I knew I had lost him when I sent him an email detailing fact-based proof that his belief in the evil power of Soros had no basis in reality and he wrote back, "Do you even check your own citations? These are all from the mainstream media." Within a week after that exchange, Jonah packed up his things and left for Indiana or Ohio to live with some people he met on the internet.

Almost immediately after the election Facebook was waking up to what had happened while they were asleep at the wheel, or rather, they were being forced to *appear* to awaken. In the aftermath of the election, the voices of academics and civil-society groups who had been ringing the alarm about features like that carousel suddenly were all over the news. It would later come out that Russia had been involved in spreading misinformation, but in a classic example of the private sector outperforming government, the much larger player in spreading misinformation turned out to be a cottage industry of people monetizing misinformation using AdSense, Google's product for hosting ads on websites.

It would later be revealed in reporting that spanned *Wired,* the BBC, the *Washington Post,* and many other sources that the secret misinformation capital of the world was the tiny town of Veles, in Macedonia. A single Macedonian entrepreneur, Mirko Ceselkoski, had set up a school to train people in how to build low-quality websites targeting America as far back as 2011. Many of his students had taken his lessons and for the 2016 cycle built out over a hundred news sites, overwhelmingly supporting Trump, which blasted misinformation across Facebook. The winning formula was simple: Step 1: Create a website that Americans believe to be a news site. Step 2: Write, or better, *steal* articles from elsewhere on the web about political topics. Don't have perfect English? Don't worry, as long as the headline is gripping. Step 3: Post it to your Facebook page with a link pointing back to your "news" site. Step 4: Watch the AdSense dollars roll in. To the residents of Veles it seemed like a miracle — high school dropouts were making more money over a few months than most Macedonians will ever have in their lifetimes. Buzzfeed was one of the earliest outlets to commit investigative journalists to dig into the malevolent innovation cluster and described it as a "digital gold rush." By 2020 we knew lots of details about how this all went down because many of the operators felt no shame about their actions. Ceselkoski let researchers and journalists visit his operation and examine his training materials in the years after the election. The "publishers" enjoyed bragging about how much money they took home.

Using data like Google Analytics or even just Facebook itself, these information entrepreneurs, or rather disinformation entrepreneurs, would watch which of their stories drove the most traffic back to their sites and then continually tweak and refine their formula for a "successful" story. Add in a little signal boosting in the form of posting to large groups, and they could drive huge quantities of traffic back to their websites to view the ads.

We used to have a gallows humor joke about the Macedonian misinformation entrepreneurs when I worked at Facebook. It would begin with confidently pointing a finger to the sky while pronouncing: "We can be confident that there will be *NO* Macedonian misinformation problem for the US 2020 election." And then a pause ... "Because Macedonia is now the *Republic of North Macedonia*!" A play on the country's 2019 name change.

The carousel disappeared soon after the election. Only after I joined Facebook did I learn that Civic Integrity, the team devoted to making sure Facebook was a positive force in society, had been spun up in a hurry in the wake of the 2016 election. They went from a very small team focused on civic health initiatives like helping people register to vote or find their polling place, to an impromptu strike force figuring out what Facebook needed to do to stanch the toxic bleeding. Samidh Chakrabarti, my future boss, had signed up for a job focused on feel-good initiatives like increasing civic participation but less than two years later found himself at the center of the maelstrom, attempting to fight a forest fire burning not just through America but around the world. But I didn't know any of that yet.

Over the course of a six-month period at Pinterest, my team on the Home Feed had shipped at least four or five changes that were each considered significant enough that we were called repeatedly on stage at Friday all-company meetings to be recognized for how much we had pushed the business forward. One of those successes removed a simple bias in our algorithm that had been holding the company back. Our blind algorithm had no concept of "boredom"—it saw no problem in presenting you with five more pins exactly like the last one you pinned. Remember, at least today, artificial intelligence is not intelligent. By intervening to increase variety with users' home feeds, we increased total content viewed every day by 6 percent. Given that Pinterest, like Facebook, was an advertising business, increasing consumption by 6 percent translated into increasing revenue by 2.5 percent. When applied over a $700 million annual baseline, a 2.5 percent increase translated into serious money.

I felt like I had finally gotten my stride back. My team had successfully migrated our whole pin-prioritization system to a new "neural net" style architecture, an improvement that increased consumption globally by 5 percent by being able to process far more data each time it decided whether to promote a pin. In one of the proudest accomplishments of my career, I led it through launch review in a single review session. The head of engineering for the Home Feed algorithms said that every previous time they had pushed through a change of that magnitude, it had taken seven or eight sessions to bring the thirty people who represented the various stakeholder groups of Pinterest into consensus—a massive productivity tax on the company. I had done it by sitting down and working with each individual stakeholder, one-to-one, to resolve their concerns; we didn't need to wait for a single raised hand to derail the process once the group convened.

I can be understanding of Facebook's loyalty to navigating by numbers because of what came next. Because the quantitative side of the company wasn't really understood or valued, the two engineers who worked on the neural net migration with me received only so-so "meets expectations" on their performance reviews. They responded by transferring to the Monetization (ads) team, which was highly focused on metrics. I wanted to join them but was unable to because in my performance review I'd received a "misses some expectations" and was reviewed as having "communication deficits." I was stuck with my manager until I could make him happy.

Right then and there I informed my manager that I was going to take my accrued vacation and quit. I gave him three weeks' notice before I hit the beach. He was floored and suggested I take some time to think it over. I probably should have — once you know you're going to quit, it's an amazing opportunity. Afraid to have that hard conversation with your boss about your relationship because you're worried it might hurt your promotion prospects? Have a project you want to pitch but are worried your boss would criticize you for not focusing on your other duties while you looked into it? Maybe there's a skill you want to learn, but are worried to allocate time to it because it might hurt your performance review? Guess what? You're already willing to quit, so why not try? Maybe I could have treated it like graduate school and just leaned in to learning more machine learning and data science skills until my boss noticed. I could have said I needed to because my engineers were gone. At a minimum I would have saved up a bigger nest egg by staying employed longer.

But I didn't take advantage of any of those opportunities. I was angry that my star engineers were gone and I couldn't follow them to greener pastures. I was tired of feeling like the work we did wasn't seen, and I had no idea how I would be able to get "meets expectations" with an engineering team that had lost at least half its capability. If what we had shipped in the previous six months wasn't enough, whatever would be? In the end, I just didn't trust my manager anymore. It was that simple. I made an emotional decision, and I quit.

I shouldn't have rushed out the door at Pinterest. Instead of taking time quitting Pinterest and using that opportunity to do a leisurely job search for an ideal next step, I boomeranged back to a start-up that had given me a competing job offer when I was negotiating with Pinterest. Because I acted impulsively, I spent much of 2018 at a start-up that ran out of money. After the company was reduced by about 30 percent, I realized I didn't have enough faith in them securing their next round of funding and decided to take some time off. It had been four and a half years since my brush with death. I had rebuilt my body, my social circle, and my confidence in my abilities. I felt like I was ready to take on my next challenge. I just didn't know what it would be.

And then Facebook called.

CHAPTER 9

Expressions of Doubt

Our deepest fear is not that we are inadequate.
Our deepest fear is that we are powerful beyond measure...
As we are liberated from our own fear,
our presence automatically liberates others.

—Marianne Williamson, *A Return to Love*

When a Facebook recruiter emailed me in late 2018, the company's brand was already sufficiently damaged that everyone in Silicon Valley above a certain seniority level was continually getting peppered with emails from Facebook recruiters. This was the era immediately after Facebook had been outed by a whistleblower, Christopher Wylie, for letting Cambridge Analytica steal the personal information of millions of people, and even Mark Zuckerberg couldn't prevent himself from being dragged in front of Congress to testify. In the end Facebook disclosed 87 million users were impacted.

I wasn't sure if I wanted to join Facebook, so with the confidence born of not caring if they shot me down, I told the recruiter I would only be interested in a role that dealt with misinformation. I didn't know whether such a position even existed. I was surprised when the recruiter followed up and told me there was an open position as a Civic Misinformation product manager. Still, I felt ambivalent enough about Facebook that it took me until the following April to actually interview. They offered me the job later that week.

For at least six weeks I dragged my feet about giving my answer. To me, taking a gig at Facebook didn't add value to your résumé; if anything, it left a dent in your personal brand. When I was at Google, the general take on Facebook was: *We are organizing the world's information and making it universally accessible and useful. What does Facebook have to offer? Your roommate from college sharing photos of their cat? Details of your cousin's breakfast in Topeka? A giant pile of equity?* May 2019 rolled around, and I was told I had to decide if I wanted to take the job or let it go.

While I considered it, I talked to some friends who did not work in tech. Thinking out loud to them, I said that I felt conflicted—something like, "I feel like I have a duty to take the job, but I also don't want to." These were friends who worked in the arts, teaching circus acrobatics to kids and adults. I felt ridiculous as soon as I said it, and the way they looked at me didn't dilute the sensation. How could anyone think they could influence a company like Facebook?

Part of what I wrestled with was that I felt that I was one of the relatively few people in tech who understood the ins and outs of social software. Not many people have worked on the recommender systems that direct the attention of billions of people around the world. I had spent years digging into how a machine could introduce unwanted patterns along the way to meeting their business goals, all the while oblivious to consequences. And I had worked on three social networks already. Did I have an obligation to step up when the opportunity presented itself?

And I had personally witnessed the consequences of what could go wrong firsthand—through Jonah. If a smart, emotionally intelligent, intellectually curious young man could disengage from reality because of where the internet pushed him, what chance did people with far fewer advantages have? If I was one of the small cohort who understood these systems, and if I could be a part of reducing, even the slightest bit, the amount of misinformation distributed, didn't that mean I had a responsibility to at least try? I thought back on the pain and powerlessness I felt when I watched Jonah separate from our consensus reality. If I could keep one person from feeling that pain, keep one person from falling into that deceptive maw, it would be worth it.

What weighed down the other side of the scale was that I had been suspicious of Facebook for years. Back in the mid-2000s at Google, defections to Facebook were rare, but they represented a nonrandom set of the people I knew. I remember one executive who had a reputation for being a bit cutthroat (which was a rare thing at Google), who left for Facebook only to bounce back to Google maybe a month later. We all assumed he had significantly increased his total comp in the process. Facebook was famous for paying a brand-impairment compensation premium—you would definitely get paid more at Facebook than at Google, Microsoft, Apple, or Amazon. When that pay bump came on top of an already elite software engineering salary, you could imagine what kind of selection pressure that might create.

Trying to make the most informed decision, I asked to meet with the person who would be my engineering peer, the engineering manager for the team. That person hadn't been hired yet (which should have been a red flag), so I met with the head of engineering for Civic Integrity. He was smart, funny, and well read. He had been a founding member of the Civic Integrity team and had risen through the ranks as it had grown. I felt like I could trust him, and I was excited about working with him in the future.

Maybe this could work. I took the job.

My first day at Facebook, June 10, 2019, began ninety minutes before I needed to be there when I climbed the stairs of a double-decker Facebook bus that stopped in the Mission District of San Francisco, just a block from my apartment. Traffic along Highway 101 had dramatically increased since the days I joined Google in 2006. Facebook is twenty-nine miles from where I lived in San Francisco, while Google's original campus is thirty-six miles away. Back in 2006, I thought the forty-five-minute one-way commute on the Google bus seemed like an unbearable amount of time. Now the commute to Facebook's office at 1 Hacker Way took seventy-five minutes, and that's if it didn't rain. Was I really willing to spend two and half hours a day on a bus to work at Facebook?

The Facebook buses gleamed and were empty. Like most tech shuttles in San Francisco, they looked nondescript. Only in the movies do tech companies paint their logos across the sides of their buses. I remember in the depths of the financial crisis in 2008 that people were throwing bricks through the windows of Google shuttles. No need to attract unnecessary attention should things go sideways.

Facebook had placed an overly optimistic order for the custom buses a few years before, expecting their staff to continue to grow exponentially. Yet due to either higher-than-expected employee churn or greater difficulty recruiting people to join, the fleet was significantly underutilized. I took two seats for myself and propped my legs up. I was still experiencing a fair amount of pain due to my neuropathy; being able to get my feet off the vibrating floor kept the nerve tingle from flaring up.

The custom-built headquarters building looked like something out of a science fiction movie. It was a quarter of a mile long, maybe six or seven stories tall, and a gray megalith. Buildings can be built for the pleasure of those who look upon them, the pleasure or utility of those inside them, or both. The blank, brutal facade of this one reflected in many ways how Facebook viewed its relationship with the world. This headquarters was a dramatic change from the company's previous corporate home. Facebook had moved into the old Sun Microsystems campus, nestled along the shore of the San Francisco Bay in the shadow of the Dumbarton Bridge and Highway 84, back in 2011. They had fundamentally reshaped the buildings, making the inner faces of the cluster of buildings into an idealized small town. What once had been boring brick facades were now Disney World–like stylized storefronts. Over the course of a five-minute walk down the main thoroughfare, which separated two rows of office buildings, you might pass an ice cream store, a bicycle repair shop, a Mexican cantina, a barbershop/hair salon, or many other "businesses" meant to improve your quality of life. I say "businesses" in quotes because, for Facebook employees, most offered free goods and services. When the company had outgrown that campus, they commissioned the monolithic headquarters building I would be spending my days in.

The new building took that isolated, inward-looking nature of the previous campus up a level. The old campus had security posts at each of the points where one might walk onto the grounds. Now the village was encased in a single easily defendable outer membrane. The only way you could ascend to the main level was by entering through one of the handful of lobbies on the ground level, swiping your keycard not once but twice before you were inside. We didn't have retina scanners, but you could imagine that someone had at least considered installing them.

My bus pulled into one of the docks where three of them could unload simultaneously. There were so many lanes for the buses in this port that at the end of the day, to reduce the confusion of so much coming and going, each lane had a person managing the flow with a megaphone and tracking clipboard. "No," they would say to comfort me, "your bus is late, you didn't miss it." One had to respect the operational excellence of it all.

The main lobby took my breath away. The atrium was massive, fifty feet high, and adorned with (genuinely good) abstract art. This was a place where entryways were cathedrals. As I reached the top of the stairs, my head swiveling in awe, I was greeted by someone from orientation and herded along with my fellow new hires.

The most important piece of career advice I give young people is, "Always remember: You can get paid in many different ways." Our culture places emphasis on high-paying jobs, but people derive value from their work via many different sources. Yes, there is money. But there's also pleasure in doing the work. Recognition from your peers, or similarly, recognition from larger society. In other words, are you cool? Admirable? Maybe you get stable, predictable income. Or you have the ability to control your work. Work/life flexibility or integration is hard to overvalue. Creative stimulation keeps you engaged. A feeling of purpose is priceless, as is giving back to others. Large tech companies pay well, but for many, start-ups might pay better and come with less legacy baggage. So for the big companies, campus architecture is one of the easiest ways to impress (and compensate) employees. Standing in this triumphal setting, my qualms about working at Facebook were calmed. In that moment, I felt like I had made the right choice. I had been through so many ups and downs in the five years since I left Google, and this was my reward for rebuilding my career.

Many large tech companies divide up the orientation process into phases. The first day pretty much all of the new hires begin the onboarding together. You pick up your access badge, here's your laptop, let's set up your health insurance. After that, the small army of new hires is broken down according to their roles. When I joined Facebook it was different from any company I had worked at before in that there were two full weeks of boot camp for all new product managers. Somewhere along the way some-one had noted that the washout rate for new product managers was extremely high. Facebook was acknowledging that it did things a little dif-ferently than most companies in Silicon Valley, differently enough that we would need a robust primer on how to be successful within the company.

However, my two weeks were cut short, by a lot. Only three days into boot camp, my manager instructed me to leave my training and pull together the team plan for the following six months. This was my first red flag that something was profoundly wrong.

We were an entirely new team: I was a new product manager, my engi-neering manager had joined six weeks earlier, and our data scientist was new. Even our engineers had been pulled off what they had spent the last few years working on: user-facing products within Facebook. Collectively, we didn't know much, if anything at all, about how Facebook's algorithmic systems worked or what the causes of misinformation were within its sys-tems. Yet we were informed that in about ten days we would have to face all the senior leadership of the Integrity organization and explain what we, as the Civic Misinformation team, planned to do to fight misinformation.

When I first came out as the Facebook whistleblower, one of the insults hurled at me by certain trolls was that *I* had censored stories about Hunter Biden's laptop. I can see why someone would think that — I was the Civic Misinformation product manager, after all, and the company had indeed decided that those stories were misinformation. But the group that decided which stories were not allowed to be distributed on Facebook was actually the core, or main, Misinformation team (I know, confusing). Where my team had six people and was new, the older, more established team had forty people, was located in the traditional Integrity department, and had been working since 2016 to build out the third-party fact-checking system that commissioned journalists to write "fact-checks" on a very small num-ber of hyper-viral stories.

Facebook was clear — they didn't want to be the arbiters of truth, but they were willing to support journalists who would provide the judgments on what was allowed to stay up or be taken down. Facebook formed a Mis-information team to build a platform that let their partner journalists see which stories were trending on it, and paid out small sums of money for each article the journalists wrote researching whether the post they had selected was true or false. Facebook would use those opinions to find similar posts and either remove them or demote them depending on how closely they matched.

Facebook claimed its third-party fact-checking program was a good-faith effort to promote information quality on its platform, but even the way they bragged about it highlighted its limitations:

> We're proud that in just three years, we've assembled a diverse network of over 50 fact-checking partners [usually nonprofits, which then hired journalists] in 40 languages around the globe — unmatched on other social platforms — that is reducing misinformation on Facebook.
>
> —Statement from Facebook, January 2020

Facebook never bragged about how many fact-checking articles were commissioned from each of their journalism partners around the world, or how much money they spent on all this. That was because the typical fact-checking partner wrote twenty to a hundred fact-checks per month, a fact we know only because journalists like Judd Legum would carefully catalog the output of Facebook's fact-checking partners. Let's imagine that all fifty partners in 2020 were doing even a thousand fact-checks per month, which is five to ten times more output than the largest participants in the program — that's still at most fifty thousand posts for the entire world of *three billion* Facebook users.

In reality, what few fact-checks that did exist were heavily tilted toward the United States. Seven of the fifty fact-checking partners in 2020 focused on the United States, leaving most other countries, if they had a fact-checker at all, with only one. Facebook did not allocate resources based on need — they allocated precious safety resources based on the fear of consequences for the company. Some unstable country teetering toward ethnic violence likely had greater need for fact-checking budgets than did the United States, but Facebook was a US corporation and they feared angering the public or government, which might in turn force oversight of their operations.

Over time, I would come to recognize that Facebook very selectively presents the data regarding its efforts to fight misinformation or any other harm on the platform. They love to share stats with the public that tell the story they want to tell, but not numbers that would actually shine light on what was happening behind closed doors. Facebook talked about third-party fact-checking like it was the solution to misinformation, but the data made clear that it could at best scratch the surface. At most, it felt like a showpiece.

But the Civic Integrity team knew exactly how much — or how little — of the misinformation problem was being addressed by those few fact-checking articles. So after years of begging the core Misinformation team to think beyond their third-party fact-checking platform and attempt to focus on how misinformation fomented ethnic violence, Civic Integrity formed a new team — our team, Civic Misinformation. Our scope was simple if terrifyingly broad: We were responsible for misinformation anywhere the third-party fact-checking program did not operate — which was most of the world. In April 2019, Poynter, a media institute, reported that Facebook's fact-checking program covered only about fifty countries (with one partner stretched thin across eighteen of them). Every other country in the world was our domain, alongside any place experiencing "times of crisis," when the fact-checkers couldn't operate fast enough to prevent mass casualties.

One of our first team meetings, if not *the* first, was with the three core team members plus a couple of Civic Integrity researchers to discuss goals and metrics. At Facebook, nothing was false until a third-party fact-checker had researched the post and written an article proclaiming it to be so, but we were responsible for misinformation only where third-party fact-checker journalists did not operate. It was an obvious chicken-and-egg problem: by Facebook's definition, *nothing* was misinformation in the arena where the Civic Misinformation team was responsible. As we discussed all of this, it became evident that we were going to have trouble defining success in a way that Facebook valued. I had missed 70 percent of the product manager boot camp, but I had clearly learned this: At Facebook, nothing mattered if it couldn't be *measured*.

My first "Oh shit" moment came within the first five minutes of that meeting. We were discussing the potential of monitoring and measuring "expressions of doubt." This is what Facebook called the comments users would leave on misleading or just plain false posts such as, "I don't think this is true" or "This is misinformation." We considered keeping track of the rate of expressions of doubt, since a decline would serve as a measure of our success. One of the Civic Integrity researchers said, "That sounds reasonable as a way to measure misinformation, but I was really struck during our user interviews in India during the lead-up to the national elections earlier this year." (The researcher was referring to the 2019 election in India, where the far-right Bharatiya Janata Party won the majority.) "Educated people," he continued, "people with master's degrees, would say things like, 'Why would someone go to the trouble of putting something false online? That sounds like *a lot* of work.'" Meaning, there were highly educated Facebook users in India who were such new arrivals to online life that they had not developed their troll radar. They hadn't encountered enough shitposters to be hypervigilant. Because they gave whatever was posted on Facebook the benefit of the doubt, especially when it was reshared to them via friends and family, they would be less inclined to question its veracity, let alone leave an "expression of doubt."

We all went quiet for a moment. The research team for Civic Integrity was a jewel within the company, and even within the tech industry as a whole. From the earliest days of Civic Integrity, Samidh Chakrabarti, one of the founders of the Civic Integrity effort, had recognized that tech that attempts to solve a social problem can often make a problem far worse through unintended consequences if the solution isn't executed in a thoughtful manner. Given the societal implications of our work, he had always allocated a far higher fraction of Civic Integrity's head count to PhDs who specialized in things like atrocity studies or political science, people who had written doctoral theses on online misinformation, than any other division within Facebook. This was a novel action inside Facebook, and Samidh deserves recognition for holding his ground and demanding it. These researchers knew what they were talking about. That moment was my first glimmer of the scope of the problem my team had to tackle.

After it was founded in Mark Zuckerberg's dorm room in 2004, Facebook began spreading across the United States. It rolled out slowly, first only at Ivy League colleges, then to a progressively broader set of universities outside of New England. In these early days, it shared the earth with many social networks that were larger and more established than itself. Within an American context, Facebook achieved dominance as *the* social network after eviscerating rivals like Myspace and Friendster, each of which, in its own time, had seemed unstoppable. As Zuckerberg grew Facebook into the largest social network in the world, he maintained an ever-present vigilance for anything that could possibly become the Facebook killer. The purchase of Instagram in 2012 for $1 billion was born of this fear. Instagram had correctly observed that taking pictures is easier than writing, thus making content creation more accessible to a broader slice of society. Zuckerberg had seen how fast Instagram was growing, and was unwilling to take any chances, no matter the sticker price.

To ensure nothing short of world dominance, Facebook had adopted a strategy of making its platform available in even the most impoverished nations, an attempt to make it impossible for competitors to emerge. By 2022 the program, termed Free Basics, served 300 million people around the world in countries like Indonesia, the Philippines, and Pakistan. If someone used Facebook's applications their data was free, but if they wanted to use anything beyond a handful of other websites elsewhere on the open web, they would have to pay for that data. Imagine an internet that offered nothing more than Facebook, Wikipedia, and maybe six other "pro-social" apps. Paying a few dollars more for data that would allow for browsing the internet beyond what free Facebook access provided might not sound like much, but in many corners of the globe, a few dollars could meaningfully change how much food was on your table.

One of the tried and true ways for a competitor to emerge and dislodge an established business is to begin by catering to lower-income consumers who are willing to accept fewer features or lower quality in exchange for a competitive price. This provides an easier entry point into a market, which enables the business to ascend, expand its market share, and gradually become more sophisticated and a direct competitor to the incumbent. Mark Zuckerberg knew the threat, because he had been that threat once, and he did not want to leave Facebook vulnerable to that type of disruption.

Even as late as 2022, in many of the countries where it operated, Free Basics provided a floor of access as many primarily used Facebook's free services. To help contextualize how out of reach is the internet access people in Western countries take for granted, a *New York Times* reporter described that her Nigerian internet service cost her $80 to $120 per month in 2021, depending on the exchange rate. This in a country where the average monthly wage was $175. For the average person, even with mobile data, it was hard to justify the pennies it might cost to access news through CNN or another independent news source, assuming that a high-quality independent news provider existed online in their language. This phenomenon was even more pronounced a few years ago, as mobile data continues to become more affordable over time.

In these countries, while Facebook was expanding its reach and fortifying itself against would-be competitors, it was also making minuscule profits or losing money. But Facebook was becoming synonymous with the internet there. To give context on how lopsided Facebook's users versus its revenues were, in the fourth quarter of 2022, Facebook made 58.77 USD per user in the United States and Canada, 17.29 USD per user in Europe, 4.61 USD in the Asia Pacific, and 3.52 per user in the "rest of the world." Best-case scenario, most of the people who were using Facebook outside the United States, Canada, Europe, and a handful of other countries were loss leaders. Facebook was making the calculated choice to lose a little money per user today in order to make money off of those users and their data in the future. You can lose a remarkable number of pennies for a long time if it lets you avoid having to pay $1 billion, let alone $10 billion, for the next Instagram that comes along. Worst-case scenario, they were Facebook's only bulwark against a Facebook killer that could not be bought. With the rise of antitrust scrutiny, many politicians in the United States and Europe had been explicit that Facebook should not be allowed to buy more companies. With the Instagram playbook off the table, the only option was to keep competitors from getting a toehold.

How did that enable the spread of misinformation? Well, in those loss-leader countries, Facebook felt it didn't have the budget to help ensure a minimum level of user safety. This became horrifyingly evident in 2016 and 2017 in Myanmar. *The Guardian* reported that among Myanmar's fifty-three million residents, less than 1 percent had internet access in 2014. But by 2016 the country appeared to have more Facebook users than any other Southeast Asian country. The article quoted a 2016 report by GSMA, the global body representing mobile operators, stating that many people in Myanmar considered Facebook the only internet entry point for information. More concerningly, because many were new arrivals to the internet, they regarded posts as news. CNN would later detail how violence fanned by Facebook directly contributed to the slaughter of at least 25,000 people and drove 700,000 from their homes in Myanmar.

Myanmar is predominantly Buddhist with a Muslim minority, the Rohingya. In 2017 the Myanmar government unleashed its security forces on a brutally violent campaign to wipe out the Rohingya. Muslims who were unable to flee were slaughtered in a campaign of ethnic cleansing. Facebook served as an echo chamber of anti-Rohingya content that pushed that campaign forward. Propagandist trolls with ties to the Myanmar military and to radical Buddhist nationalist groups inundated Facebook with anti-Muslim content, falsely promoting the notion that Muslims were planning a takeover. Posts routinely, relentlessly expressed comparisons between Muslims and animals and advocated to "remove" the "whole race." According to the flood of hateful, racist, and patently untrue "information" shared thousands upon thousands of times over, "Time is ticking." And it all started at the very top of Myanmar's military and government. On his Facebook page in 2017, the most senior official in Myanmar's military, Min Aung Hlaing, posted: "We openly declare that absolutely, our country has no Rohingya race." After the bloodshed, Min Aung Hlaing took control of the country in a coup in February 2021.

The *New York Times* reported that as far back as 2000, Myanmar's military had sent officers to Russia to study how Russia ran its online influence operations — i.e., misinformation. Back in 2000 this involved blogs, but many of the techniques were similar to how social media would later be weaponized: First build an audience, then use that channel to spread mis-information that supports your goals.

The same Civic Integrity researcher who provided my team with the disturbing context for why measuring "expressions of doubt" might be less useful than we hoped continued to lay out what our small, brand-new Civic Misinformation team was up against. Because so many people were coming online so quickly in many impressionable and fragile markets, there hadn't been time yet for the local communities and cultures to estab-lish norms about what should or shouldn't be trusted. It was one thing to have people with master's degrees who hadn't yet experienced enough trolls and shitposters to understand what trolls and shitposters are and how they can wreak havoc on the internet, but it was an entirely different situation in places like Myanmar, where the education system emphasizes rote learning, memorization, and regurgitation. The number one defense against misin-formation is critical thinking skills.

Almost two years later, when I would begin to document what was known about Myanmar within Facebook, I would learn that 50 percent of all messages sent on Facebook Messenger were sent via voice clips, because much of the population could not write effectively. I would learn that Face-book user-experience designers were reluctant to redesign the icons used on buttons because a meaningful fraction of users of Facebook's products globally navigated them by memorizing the meaning of the images. Yes, your new, upgraded visual style would look cleaner and more aesthetically pleasing, but it would also be particularly disorienting if you didn't have a way to learn what the new images in a different layout meant.

As the internet expanded around the world, in each new place it touched, it brought with it a rapid expansion in the amount and kinds of information people encounter in their daily lives. In most places that Face-book expanded to, the average person might only have had access to heav-ily centralized media like government-run (or blessed) TV or radio stations. Suddenly, anyone could become a broadcaster.

There are a handful of moments in world history when a population has experienced such a phase change. The rise of the printing press in the late 1400s and early 1500s made it possible to produce cheap, easily distributable pamphlets. Part of what made Martin Luther's Protestant Reformation so impactful was that he and those inspired by him coached the few who were literate on how to spread the information he published. His was the first "information operation." Unfortunately, many people with less noble goals mimicked his methods and in the process unleashed the first wave of violence fueled in large part by misinformation. Pamphlets encouraged citizenry to identify and burn witches; pamphlets (some of them by Luther himself) promoted killing Jews. Changing the information ecosystem overnight without giving the world time to adapt could be catastrophic.

The same thing was happening now. Facebook made social-network infrastructure available to large communities with low literacy rates, effectively turning on a fire hose of information and disinformation, and those communities weren't ready for it. When I worked on Google+ in 2011, there were still a number of countries in the world in which Facebook was not the most popular form of social media. By 2019, Facebook and its products were the most dominant social media in every country in the world, except for a few where it was banned, like China. TikTok, largely credited as the most likely platform to replace Facebook, in many ways validates Facebook's aggressive expansion strategy into fragile markets — it arose from the only place Facebook wasn't allowed to play: China.

After our initial meetings with the Civic Integrity researcher, I felt I needed to sit down with my boss, Samidh Chakrabarti.

Samidh was a busy person. He had at least twenty people reporting directly to him by the time I was added to that roster, and I met with him for the first time around that same day. I did not have an assigned desk yet, so he told me to look his desk up in the company directory and navigate my way across the building to it. Just as the facade of Facebook's HQ showed how little they cared about contributing to the aesthetic built environment around the office, the HQ's interior implicitly reflected what they viewed as the ideal corporate culture. As I walked across the building, I could only be in awe of the scale of it. It was a quarter of a mile long, and the main office space was basically one room at least three stories tall, minus a scattering of conference rooms and bathrooms. A five-thousand-person open-floor office. No one sat above or below anyone else; we were all equal — just some more equal than others.

In orientation it was explained to us that Facebook's focus on metrics for evaluating everything was grounded in the idea of freedom. Most companies that get to be Facebook's size ossify because they create more and more layers of management. In most firms, if a frontline worker has a brilliant idea, they have to convince people above them — people who might have vested interests in how the status quo operates — that they should have permission to try out that idea. Facebook understood that social media is a young person's game — if they wanted to make products that appealed to young people, they had to give freshly graduated workers the ability to meaningfully change the product. Facebook believed that if they set the right goal metrics, they could let everyone run free as long as they pushed those goal metrics up. Everyone, and everyone's ideas, were equal. The ginormous single story open-floor plan was that corporate culture made manifest.

And it was hard to say it didn't work. I remember the product manager they brought in to testify about how liberating this philosophy was. Her team had decided they were going to rewrite the payments architecture for Facebook because the one Facebook was using at the time couldn't adequately scale. They were told not to do it, but they went ahead. A quarter went by and they again were told to stop. Another quarter went by and they were told that at some point they would need to be put on a performance plan. But in the end, they pulled their new payments system out of the fire, and its stability and ability to scale was considered a pivotal piece of what made Facebook Marketplace possible. I don't know it for certain, but I'm pretty sure she was trotted out every two weeks to explain to us the importance of making big bets and dreaming big. She was proof we could do this too because Facebook was "still a start-up," where only (measurable) impact mattered.

As I made my way across that airplane hangar of an office, I couldn't avoid noticing that while this multi-hundred-million-dollar temple to hierarchical flatness showed their commitment to their mythical egalitarian ideals, it also was a symbol of the gap between dogmatism and reality — it lacked a genuinely flat floor. Over the preceding five years I had been on an extensive journey to relearn to walk, and while I could now balance on one foot or leg-press a multiple of my body weight, I still did not have healthy enough nerves that I could raise my toes as far as the average person can when I would take a step. When you build a quarter-mile-long room, you have to accept that you are essentially building an aircraft hangar, and your flat floor is only flat within a certain degree of tolerance. Every single day I worked at HQ I tripped at least once. The toe edge of my shoe catching on the uneven surface provided a tangible reminder that I was in a space that did not welcome me.

One advantage of having an aircraft hangar for an office is that you can do remarkable things with those high ceilings. As I approached the Civic Integrity pod, I was struck by the imposing black flying wing aircraft suspended above our desks that was easily fifty feet across. Project Aquila had set out to build autonomous drones that would circle the Earth providing low-cost internet access. It seemed fitting that the totem that hung over our heads as we worked on Civic Integrity, a team whose mission was devoted to making sure Facebook was a positive force in societies and elections around the world (or perhaps more accurately to clean up after a Facebook that had gotten out over its skis and wasn't fulfilling that mission), was literally the corpse of a project whose hubris was breathtaking.

Building planes is hard. Building self-flying planes that are so reliable that thousands, if not tens of thousands, of them can circle the world twenty-four hours a day, 365 days a year, for the foreseeable future, without crashing, is almost impossible. I am not an avionics expert or even an enthusiast, but it was obvious to me that was the safety bar they needed to hit to make governments and people around the world feel safe enough to let them keep flying over their communities. Somehow Facebook, a software company, thought they could hit that bar of aeronautical and operational excellence when they greenlit the project. One of the first three drones crashed catastrophically early on, and the project was cut loose. Just as Google had hung a copy of SpaceShipOne in the Googleplex offices, one of the remaining drones was used to decorate our office and inspire us. It's just a guess, but I don't think they intended for me to draw a message about hubris from it. I found it fitting a few months later when someone asked me if I had seen it yet—he said he had gone months working at Facebook before noticing it.

I found Samidh Chakrabarti's desk and we headed up to the roof to walk the meticulously manicured gardens, unquestionably my favorite place on the Facebook campus. They overlook the marshlands that rim the San Francisco Bay, and even when cold and windy, they provide an escape from work on Facebook's virtual world and a reminder of what is real. Samidh walked with me and laid out his expectations. We had both worked at Google before working at Facebook, and he emphasized to me that Facebook had a *doing,* not a planning, culture. He needed me to get up to speed fast, and he encouraged me to find even a small change to ship in my first two weeks. This was an exercise required of every engineer in the company to get them exposure to the process of launching changes. I was not a software engineer, but he thought the practice educational. He suggested I look at the Trusted Partners reporting interface—he knew there were lots of features partners wanted added to it. There must be some low-hanging fruit there. This would be my first failure at Facebook, and would come to be symptomatic of the gap between Samidh's expectations for me and the reality of my job for the next six months.

The Trusted Partners inbox emerged as a direct result of the United Nations report on the role of social media, and Facebook in particular, in the Myanmar genocide. The UN had recommended that Facebook work with outside agencies and experts to conceive and implement policies and features that would help prevent the dissemination of toxic misinformation of the sort that had led to the atrocities. Today, Trusted Partners includes more than four hundred nongovernmental organizations: humanitarian agencies; researchers from more than a hundred countries around the globe; and entities like Tech4Peace in Iraq and Defy Hate Now in South Sudan. Back in 2019, it was only operating in Myanmar.

The UN had found that in the months before the genocide, civil-society groups tracking the escalating tensions had repeatedly tried to reach Facebook with information about what was unfolding on the ground and the impact Facebook was having, but they were unable to get through. On the Facebook site itself there was no pull-down menu, no mechanism in the Get Help interface to report "imminent ethnic violence facilitated by social media." The Facebook customer support organization for the region was a streamlined machine. *Wired* magazine reported that "At the time, the company had just one Burmese speaker based in Dublin, Ireland, to review Burmese language content flagged as problematic." There was no way to get such warnings from the affected communities to anywhere near the Facebook leadership in Menlo Park. When that lone person began trying to raise the flag that things had gone seriously wrong, the top-down bureaucracy of Facebook's operational teams, the factory of actual people who interface with users, filtered out the message long before it reached Menlo Park. Remember, Myanmar was a Free Basics country — Facebook was laying out large sums of money to provide free mobile data, and it's highly unlikely they saw an equivalent return in advertising dollars. They were a loss leader for that glorious future when they would bring in more revenue than they cost.

In a show of good faith, in response to the UN findings and the recommendations of the Trusted Partners themselves, Facebook had built a parallel messaging system into the Facebook mobile app that was intended to provide direct access for civil-society groups on the ground to report concerning trends. Unlike the normal support inbox, one that you or I might use to complain about someone impersonating our account or an online harasser, the Trusted Partners mailbox was hidden in an unobvious place in the app and could be flushed clean with a single click if, say, a hostile governmental official wanted to search your phone.

Back at my desk, I jumped on a video call to New York with one of the team's engineers and we talked through what might be needed to push out a quick bug fix on the Trusted Partners inbox. She then reached out to people who had worked on the code at some point in the past. No worries, we all concluded; we could definitely ship a change in two weeks. I reported the good news to Samidh.

A day or two later, however, we discovered there was no chance we could ship out even the smallest change to the Trusted Partners inbox in that time frame. The code's previous keepers weren't aware that the code base had been neglected for so long that making quick tweaks that would be in harmony with the rest of the app was no longer possible. That particular code was like a piece of a puzzle that no longer existed. The only way you can ship an interface change to hundreds or thousands of versions of phones and have them display your change in a predictable way was to use a computer program that abstracted out your code to something a higher level up. This "framework" would then transform the different puzzle pieces you had used to compose your user interface into the actual lines of code that would make up those hundreds or thousands of different versions of your app. In the same way that the Facebook app gets new features and versions, these frameworks also similarly evolve over time, sometimes to the point where interfaces must be completely rewritten using the new components.

Engineers like making new things, not copying old things. No one wants to take the time to translate an already working product into a new framework, and so deadlines are routinely set so that if your corner of Facebook hasn't migrated onto the new standard as of a certain date, you're no longer allowed to ship changes until you rewrite it. This allows low-touch, slowly moving parts of the application to continue operating "as is" even if the owners aren't willing to bring it up to date. Facebook had promised to build the Trusted Partners messaging system when the UN shamed it, but they hadn't promised to maintain it. Soon after rolling it out, they realized they didn't have a way to effectively scale or manage the program. Given that this was meant to be a critical information channel in life-threatening circumstances, Facebook had committed to a twenty-four-hour service level agreement. If a partner wrote in with a concern, Facebook promised that the relevant team would receive it and get back to the reporter within twenty-four hours. This meant there always had to be people staffed to it, and likely surplus capacity to cover a spike of cases. This was extremely high-touch expensive support compared to how Facebook usually conducted business.

Facebook had no rewards for maintenance, and the Trusted Partners inbox was a cost center. The more reports that were filed just meant more money spent responding to them. Like many things in Civic Integrity, there really were not good metrics of success for this service. As a result, after the "touchdown" of the initial Trusted Partners inbox, some or all of the team that had conceived it had rotated elsewhere, and it was not maintained. I had to return to Samidh and tell him I would not be shipping a feature in the first two weeks, and that if we wanted to change the Trusted Partners inbox we would need to invest three to six months in rewriting the whole product from scratch.

A few days later I arrived for a meeting to review annual goals and heard something I would soon hear often at Facebook — "I don't understand why your team exists." Most often when I heard this sentiment it came from someone on the core (big) Misinformation team. I knew I was in deep trouble when the first conversation I had with the head of this team, the man Samidh said was supposed to coach and onboard me onto Facebook's systems and process and work with me on Misinformation, said not only, "I don't understand why your team exists," but also "I don't think your team *should* exist." Why should he go to the trouble to invest time in getting me up to speed to work on a project when the very existence of our team was a reminder that his team wasn't handling the challenges it should? In reality, there were so many challenges mounting so quickly that no one team could have ever handled them all.

If we had shown up at that review meeting with clear metrics to propose, we might at least have bought ourselves some time. But only weeks into the job, with an entirely new team, we didn't yet have clear success metrics — or even a good definition of success. Samidh ended up at a whiteboard arguing with the head of Integrity, Guy Rosen, drawing Venn diagrams of what the core Misinformation team would be responsible for and what the Civic Integrity team would be responsible for.

I don't want to bore you with a blow-by-blow recounting of the next six months, but I will say it did not end well. While the problem of misinformation that was beyond what could be covered by third-party fact-checkers was important enough for Civic Integrity workers to fight to build a team, an even larger misinformation problem came to light, and we were pulled off the projects we had spent the previous few months scoping and getting off the ground.

In mid-September, almost three months to the day after I joined Facebook, the Information Operations Threat Researcher team discovered a large Russia-orchestrated influence operation that was at work in the United States, sending misinformation to African American activists, environmental activists, gay activists, and police officers. In the moment, the Facebook leadership had read these groupings of targets as Russia investing in "narrowcast" misinformation for specific demographics — that is, misinformation that is tailored and sent to a specific subpopulation. Though I shared that interpretation at the time, I would later work on a project that changed my perspective. Almost certainly they were targeting different sectors of American life. Environmental activists are trusted by certain Americans more than, say, politicians. Similarly, police officers in many (but definitely not all) places are seen as trustworthy because of their work protecting the public. If Russia could introduce misinformation to these influential people and they shared it onward, recipients would give it the benefit of the doubt.

Narrowcast misinformation had been a known influence operation tactic since at least 2016. Back then, the forty-person core Misinformation team had considered it a project too hard to tackle, and more or less resigned itself to the notion that there wasn't much that could be done. In that 2016 election, Facebook had been caught off guard, and by the time they knew it, it was too late. At least, that's what they told themselves. Now, four years later, the fact that misinformation was being routed to people with guns made narrowcast misinformation come across as too high and volatile a risk not to address.

My team's mission was to address misinformation that was beyond the scope of third-party fact-checking, and that put narrowcast misinformation right in our laps. The very nature of narrowcast misinformation was of "low-reach," highly targeted misinformation. Third-party fact-checking could at most check thousands of stories a month; they could only focus on stories that might reach tens or hundreds of millions of people.

My team was informed that we needed to figure out a way to "solve" narrowcast misinformation by the end of December 2019. It didn't matter that the core Misinformation team had spent four years avoiding solving this problem because it was too hard — we had maybe two weeks to come up with a plan. This task would be our contribution to the 2020 Lockdown, which was what the Facebook Integrity bosses labeled the effort to prepare and fortify the company for the November 2020 election. Up until that point, Facebook had faced only a handful of lockdowns, each for a problem viewed by Facebook as existential. One lockdown was to build their first mobile app. Another was to bring their Android app up to an acceptable level of quality, an existential issue given that most global users of Facebook use Android. As part of a repeated theme inside the company, it had been a nearly invisible problem to Facebook's almost exclusively Apple-iPhone-using engineers until it reached lockdown levels of crisis.

The 2020 Lockdown was Facebook's admission of how far behind they were when it came to being prepared for the November election. Around the same time the Russian influence operation was detected, a group of maybe fifty leaders and specialists across the company came together to "Red Team" the 2020 election. Often in war games, a Red Team attacks and a Blue Team defends. In this case, it was a coincidence that the Red Team was coming at Facebook Blue (what facebook.com is actually called within Facebook). A Menlo Park conference room became mission control for the war game, and for hours it buzzed with people throwing out all kinds of scenarios in which operatives, trolls, and psyops operations could compromise the 2020 election.

The result of the exercise was a Risk Grid with colored squares that represented how ready (or not) Facebook was. It had ten problem areas represented as rows — things like information operations, misinformation, voter suppression, hate speech, account security, and harassment. Across the top were maybe six or seven columns as surfaces like Facebook Blue, Instagram, Whatsapp, Ads, Groups, Messaging. The intersection of the rows and columns represented a problem to be solved. For example, voter suppression (row) on Instagram (column).

The election was a little over a year away, and going into the war game the thinking was that there was still plenty of time to course-correct. When you ran your eyes over the grid, you had to ask yourself whether perhaps the exercise should have been run eighteen months or two years before the election instead of twelve. Of the sixty squares in the grid, all but a handful were red (meaning highly problematic and vulnerable), several were yellow (meaning not ideal, but not critical), and almost none were green (meaning we're good to go). The point of the lockdown was to detect and zero in on high-priority projects. My manager wrote in a post when the lockdown was announced, "Lockdowns are not about 'working more,' they're about accelerating progress by creating extreme clarity on a small number of the most important things." Teams like mine would be pulled off existing tasks to address such projects.

Yet after the war games had concluded, there were roughly sixty priority projects. Even after grouping some of the squares because a single solution might work across multiple platforms, there were still an unmanageable number to accomplish. Now the leadership created a darker, dried-blood shade of red to represent the ten projects that were considered the highest-severity risks to the election. It was as if we had war-gamed a nuclear confrontation with Russia and China, and fifty of our scenarios had ended in total annihilation. We felt forced to choose ten of those fifty where the annihilation was just a bit more gruesome. At Facebook you learned to look at those scenarios and say, "We just don't have the resources" to stop fifty different paths to war. And then you didn't follow up by asking *why* we didn't have enough people, or whether we should be the ones allocating the engineers. If the public had known which squares we had left red, would Facebook have chosen to keep developing rarely used features in Facebook ads over turning more squares on the 2020 Risk Grid green?

After a crash course in the problem space, we could understand why the core Misinformation team had not touched narrowcasting. Third-party fact-checkers were only interested in the most viral misinformation. Given the tools Facebook provided, they would never even notice a false story circulating only among police officers, let alone choose to review it.

Instead of focusing on the "misinformation" part of narrowcast misinformation, we focused on what it meant to "narrowcast." People narrowcast every day when they communicate to a community via a Facebook Group—that clearly wasn't the problem the threat researchers had found. We decided the problem we were trying to solve was when people were trying to aggressively communicate to a discrete community of people. Our bad actors weren't sending a post a day to their cancer support group, they were likely sending tens or hundreds of messages to evangelical Christians or older black men or, for that matter, police officers, as part of an organized campaign. The "problem" to be solved was finding those overactive, targeted sharers.

To find people who target discrete communities, you have to have a way of dividing up the United States (or another country) into subcommunities. This sounds simple. We have well-established taxonomies, ways of naming, describing, and classifying people that marketers, politicians, and even you and I use every day. We might divide people by age or by gender, maybe by race or ethnicity. But while some descriptions are objective, like how old a person is, other categories are more fraught. Should Facebook try to guess if you are black? Should Facebook guess if you are gay? That might sound innocuous, but in some countries people can go to jail for being gay.

We had no way of knowing anything about any user for sure, we could only write computer algorithms to guess whether you were exhibiting patterns similar to those of other people we thought were in a specific category. We decided that in order to respect the privacy of Facebook's users, we would develop a system that grouped people into subpopulations that had no names, just identification numbers. These groups would be coherent (the users within them similar to each other along a number of dimensions), but we intentionally never tried to affix a label to any given group or tried to understand what specific factors made the group coherent.

This way of thinking was a legacy of when I worked at Pinterest and founded the group that built the first skin-tone filters for Search. Before I moved to Puerto Rico in 2021, maybe six months before I came forward, I was a shade of white that I liked to describe as "jellyfish white." I felt frustrated that I couldn't filter clothes on Pinterest that were appropriate for me and the other vampires who roam the earth. I had joined forces with a redhead who had similar problems, and we built out a system to help users work around the natural bias of Pinterest's search engine toward prioritizing content from tanned light-skinned users. The majority of Pinterest's users in 2016 were tanned white people. An AI trained to maximize repinning that doesn't know that skin tone exists (or matters) will bias toward pins depicting tanned white people for queries like [eye shadow] because on average, that pin will perform better than a pin depicting either a jellyfish person or a person of color — because on average the user searching for eye shadow is another tanned white person. When diverse content doesn't appear in search, it discourages content creators like makeup brands from creating diverse content.

Our feeling was that we should not decide that you are black, because the world is complicated. There are lots of people who might exhibit patterns of searching that are counterintuitive for who they are. Maybe you're a white mother with an adopted child of another race. Maybe you're excited about Japanese fashion but are not Japanese. We didn't feel it was our place to put a label on people using Pinterest.

When it came to narrowcast misinformation, once you had a way of dividing up the United States, you could easily find people who were outliers in how much they were posting to Facebook, especially if they were laser-focused on only one or two subpopulations. It would be even easier if those two subpopulations were an improbable combination, say, targeting Central Florida grandmothers and Pacific Northwest hipsters. Combined with other tooling already available at Facebook, this seemed like a solution that might be good enough to make a dent.

We were clear: This was going to be a heavy lift with the team as it was. By that point, we were down to effectively four people on the project. We had acquired a core data scientist, but lost two of our front-end engineers in the constant shuffling of Facebook, and our engineering manager's skill set wasn't with data. We were extremely clear up front: We couldn't commit to measurement this quarter unless they gave us more people. They were only getting a proof-of-concept prototype. Still, we got the green light. The Russians were again looming, and I think the Integrity bosses would have said yes to any reasonable plan.

Around mid-September 2019, the same time as Lockdown motivated reorganization of the work on Civic Integrity, one of the most important projects the larger team had undertaken was drawing to a close. Thirty-plus people across the company had been guided through a process to determine what were the conditions under which Facebook should step in and take down speech from political actors. It had arisen from necessity. Earlier in the spring the Bharatiya Janata Party, the ruling party in India, had started posting materials on Facebook describing Muslims as rodents. The Civic Integrity research team immediately flagged it as an issue of concern. One of the most important warning signs for ethnic violence is when political leaders start comparing a minority to pests. It's a critical step in the dehumanization and "othering" process that desensitizes people to the suffering of their neighbors.

A proposal was quickly escalated for Sheryl Sandberg's review. Sheryl had joined Facebook in 2008 as the Chief Operating Officer and Mark Zuckerberg's right-hand woman. She had agreed it was necessary to take down the posts, but then things went astray. Sheryl had to step out of the discussion for fifteen minutes to attend to other business, and in that time someone raised the elephant in the room. If they took down speech from the ruling party in India comparing minorities to pestilence, they were also going to have to take down speech by Donald Trump when he inevitably demeaned ethnic minorities in the United States using similar language. In classic Facebook fashion, they kicked off a process so they could come up with an airtight defense to justify why they were going to take down such speech and when.

That process had finished the week before, in a review with Zuckerberg on Thursday or Friday. I was never able to read the original, but I guarantee it was specific and limited in scope, given the sensitivity. If it was anything like the many other Facebook policy documents I encountered over my two years there, it likely provided detailed criteria for what counted as speech risking communal violence, how to establish what the baseline risk was in a community or nation, and progressive tiers of appropriate interventions.

On Civic Integrity, we all expected a green light — after all, we're talking about a relatively narrow scope of content that was inciting violence. Something like a political speech policy has implications involving many different Facebook departments, from advertising to government affairs to public relations to the Social Cohesion (i.e., communal violence) team itself. Months had been invested in building consensus among such a wide array of stakeholders.

It was such a large project that many who were not on the working group were still aware of it, and we were all floored when Zuckerberg announced that the policy drafted by the working group was insufficient.

He would write it himself. Over the weekend. Mark Zuckerberg could set such a tight deadline because his policy was simple: Facebook would not touch speech by any politician, under any circumstances.

Symptomatic of the development process that created it, the public announcement of Mark's policy decree had been so rushed that no one had even consulted with the Advertising or Misinformation teams, which were now among many groups left to deal with a nightmare situation. There were hundreds of thousands if not millions of politicians, not just in the United States but worldwide. How was it possible to keep track of them all to exempt them from moderation policies? For every national-level politician in America alone, there are many more city council members, all of which, according to Mark, enjoyed the same level of protection as the president. Did the municipal dogcatcher count as a politician if it was an elected post?

Everyone on the Civic Integrity team looked shell-shocked. Samidh called an emergency all-hands meeting to clear the air. He began by telling the story of his own journey on Civic Integrity. At one point he flashed up MRI images of his spine to illustrate how much he, just like the rest of us, had sacrificed to pursue the mission of Civic Integrity: He had spent so much time sitting and working in the wake of 2016 that he had ruptured at least one if not two discs in his spine. He talked about missing time with his children during those priceless years when they were small, and how sometimes the cost felt like it was just too much to bear. This went on for at least twenty minutes. I can't speak for the room, but I certainly thought the only way this monologue could end was with Samidh stepping down.

Stepping down would not have been a rash act, either—Civic Integrity had worked according to the Facebook process. The group had diligently wrangled vastly competing internal and external interests into a brokered compromise. It showed remarkable disrespect to think you could spend a weekend alone on your computer and do it better.

And then Samidh swerved and said that all of that didn't matter, though, because we were working on something so important the sacrifice was worth it, and he was going to continue to fight on.

This kind of sentiment unquestionably contributed to the mental health challenges on the team. Every person on Civic Integrity who had worked on an issue related to communal violence was locked in an unforgiving tradeoff. Now that you were at Facebook, inside the curtain that shielded the public from the truth, you knew exactly how bad things were on the platform and Facebook's role in it. But you also knew that the work to blunt those damages was underfunded and lacked institutional will. If you stayed, you knew there was at least one more person trying to keep the train from jumping the tracks. You might have your doubts about whether what you contributed to preventing disaster would be enough, but at least you were trying. If you left, you knew there was one fewer person trying to buy enough time for the better angels of men to prevail.

A large part of what would push me over the edge into whistleblowing was that I watched that formula play out tragically over and over again. People would drive themselves into the depths of burnout trying to fight the institutional decay because Facebook couldn't heal itself with the system of incentives that existed. I think that witnessing scenarios where people's lives were on the line but too little action was taken also exposed many to what is known as "moral injury." The US Department of Veterans Affairs explains moral injury in this way: "In traumatic or unusually stressful circumstances, people may perpetrate, fail to prevent, or witness events that contradict deeply held moral beliefs and expectations" and states that "Moral injury is the distressing psychological, behavioral, social, and sometimes spiritual aftermath of exposure to such events."

I remember when I first joined the team and was thrown off by the flat affect I regularly encountered among my peers within Civic Integrity. Both positive and negative emotions seemed to pass through a filter before they reached their faces. As I witnessed moments like the establishment of the political speech policy or moments when hard-won safety changes were rolled back for what I perceived as flimsy reasons, I began to wonder if what I witnessed among my friends and colleagues was some form of moral injury, only one so prevalent across the team that it faded into the background for most of them. Civic Integrity was full of kind, conscientious, motivated people who wanted to make Facebook a positive force in the world. I can't imagine what it felt like to have months of your work — work that you thought was needed to prevent atrocities — crumpled up and thrown in a trash can.

Throughout the fall I found myself agonizing about whether I was just providing window dressing rather than dealing with misinformation on the platform. As I learned more about Facebook, it became clear to me why my job was available when the recruiter and I began talking in January and why it stayed open for at least four or five months while I was trying to decide whether to interview and later join. Almost all hiring managers will prefer a current company employee for a job before considering people outside the corporation. Someone with a track record of operating within the cultural constraints and available resources of a firm is a much less risky hire than someone entering from a potentially radically different context. My job was offered externally because no one internally wanted it. Those who knew anything about how Facebook worked, what it prioritized, or even how they defined "misinformation" knew this was a bad job. My job stayed open because the few people who had the skills for the problem space anticipated it wouldn't be worth the effort.

I should have recognized my hubris in thinking I could make a difference maybe two months into Facebook, before the pivot to narrowcasting, when the onboarding mentor I was assigned told me I was working on a problem that would not succeed at Facebook. He was a long-term veteran of the company's product management org and viewed as one of the most effective non-executives in the company.

It was still early on, and he had asked me about our problem space and our success metrics and immediately told me, "You need a different problem. This is a bad problem—it isn't measurable." He was referring to the fact that nothing at Facebook was false unless a journalist from a third-party fact-checking organization wrote an article declaring it so. And my team only dealt with misinformation outside the realm of third-party fact-checkers. He wasn't wrong.

There is a style of communication known as seek to understand. This stands in contrast to seek to persuade, where you're communicating in order to convince someone of an idea. I did myself a disservice in that moment by letting my shock influence my conversational style. I should have asked follow-up questions about what are good and bad problems at

Facebook or what issues he foresaw arising if I continued working on Civic Misinformation. It would have saved me plenty of anguish. Instead, I talked to him about all the terrifying things I had learned in the previous few weeks. About the civil-society groups telling Facebook that images were being repurposed—copied from massacres years earlier and presented as if they were contemporaneous—and used to fan ethnic violence in African countries. About the Facebook Policy team having to wait until people were actually dying before toning down how incendiary Facebook's content distribution algorithms were. About frontline NGOs having no way to tell Facebook about emerging crises it was igniting. I tried to explain the stakes, the problems that would not be addressed if our three projects weren't done. I didn't get that I was fighting the ocean.

In Civic Integrity, many of the problems we worked on were not easily measurable. Measurement is difficult, especially when you're at the start of a problem and you have to define the very space you're operating in. For the first few years, Facebook gave Civic Integrity the space to run in headfirst and just start battling the forest fire. With time, it was expected that Civic Integrity would come into alignment with how the larger Integrity organization operated.

Sometime during the US 2020 lockdown, I assume Samidh's manager began exerting pressure for that transition to begin. We had stated clearly during our start-of-lockdown review that we could not do measurement and build a prototype within three months. If they wanted both, they needed to give us more engineers. No problem, they said. But about four weeks later Samidh was holding twice-a-week reviews with us covering only one topic: measurement. We suggested using the Misinformation Classifier, a computer program built by the core Misinformation team that predicted whether something would be assessed as "false" if a journalist fact-checked it. He said it wasn't enough, and shared an internal-to-Facebook article that explained that using classifiers for measurement was bad because teams would begin to "teach to the test." They wouldn't purely solve the problem, they would focus on the content the classifier could detect. We needed "real" humans assessing whether our system could find misinformation.

Instead of building our prototype, we diverted to research available

Facebook tools and vet which of the various pools of reviewers located across the company we might use for evaluating our system. Three weeks later, we had successfully walked through a full round of evaluation of whether what we had found with our system was misinformation. Everything we did during those three weeks was irrelevant — the code we wrote would not be appropriate for our final prototype or how it would be used — but now we could say we could "do evaluation." And Samidh could report to whomever was clearly bearing down on him that one less project in Civic Integrity was measurement-free.

It was almost eighteen months later when I realized the true irony of those three weeks. I was in the midst of gathering information on how Facebook operated when I stumbled on the article my manager had waved at us to prove he was right. It had a dramatic headline, something like, "The Dangers of Using Classifiers for Measurement," but it came to a conclusion that was the opposite of what he had described to us. Sometimes problems can't be evaluated easily *except by* classifiers. To keep from teaching to the test, you need to keep updating your classifiers to be more accurate, so the team being evaluated will keep running along behind you. We had tried to explain that part of the danger of narrowcast misinformation was that it was hidden — because narrowcast misinformation reached fewer people, journalists out in the wild might never see the falsehood and get to report that it was wrong. The results that came back from our test evaluation round corroborated this. We had used a team that would go look for validation already published on the internet, and at least half of the judgments came back as ambiguous: "Could not verify if true or false." The machine learning classifier took a different strategy. Instead of looking for factual support as a reporter might do, it drew on differences between how journalism and misinformation differed stylistically. For example, legitimate journalism plays less often on fear and uses less sensational language, among other signals. Our wild goose chase truly was for nothing.

The whole incident was symptomatic of a larger trend that prevailed throughout the first six months I worked at Facebook. At every previous company I had worked at, for any team I worked on, the engineering manager and their boss always had experience working with algorithmic systems and data. Even my product management people-managers had hands-on experience working with algorithms, or at least enough experience working with algorithmic products to know they should defer to people who did understand them.

The technical staff of Civic Integrity was staffed by the willing, not necessarily by those with the most relevant skills. It had grown organically, staffed by people willing to make significant personal sacrifices to attempt to protect the public, and it had a tradition of tackling problems with whomever was on hand. The composition of my team was representative of that strategy. If I had been in Samidh's shoes, I would have hired a product manager and paired them with a data scientist and a researcher. After a month (or maybe two), I would have asked them to describe what they viewed as the most critical problems after they had dug into the data and consulted relevant teams with adjacent exposure to the space. Only then would I have placed engineers on the team.

Engineers are like doctors in that each field of practice has a wide range of specialties. You don't go to an ophthalmologist about a problem with your foot, and you don't go to a podiatrist about a problem with your eye. Neither is better or worse than the other — they each have their own important domains — but you will deeply regret relying on the wrong one to help you.

My team was formed backward. Instead of assigning an engineering manager and engineers most appropriate to the problems we were trying to solve, I was told I should be "growth minded" and coach the engineers I had into being able to solve the projects we found. And fast, because the worst thing an engineering manager can do is have engineers sitting around doing nothing. Instead of taking the time to find the right projects and recruit the right team internally or externally, I was pressured into saying yes to projects our engineers were unlikely to be able to complete.

I was working with three front-end engineers, experts on how to put together the user interfaces that people directly see and interact with. They were unquestionably world class and skilled in techniques that optimize for things like reliability, performance (interface speed), and aesthetics. It is incredibly difficult to build an interface like Facebook's that can be success-fully deployed on hundreds if not thousands of varieties of phones around the world, accessed 24 hours a day, 365 days a year. They weren't slouches in any way. But we were a team assigned to deal with information-based problems. If front-end engineers are experts at the visual — what can be seen and touched — it was as if we were asking them to operate in a dark and abstract world of intangible patterns. We needed engineers with experience in the trenches of data science and machine learning. While front-end engineers could certainly be retrained to operate effectively in that context, it would take at least a year to do so, and that assumed we had qualified mentors for them to work with. I am in no way that mentor.

My concerns fell on deaf ears. I found myself in situation after situation where because my engineering manager and his manager had not built algorithmic systems, they didn't know how to assess the credibility of what I said. They had been hearing about machine learning and data science for years. In their minds there was no way these things could be as hard as the hype had it. I must be the problem — I was just too negative.

By the time December rolled around, I was getting worn down trying to pull the train back onto the tracks. Another situation emerged that I should have pushed harder to stop, but I knew it would only alienate my engineering team if I pushed back too hard. They saw this opportunity as the kind of boondoggle they were entitled to as employees of a trillion-dollar corporation. A veteran Civic Integrity core data scientist, Sasha, had taken an interest in narrowcasting and had invited our team to New Zealand to cowork for a week. This trip, if it were to occur, would take place the same week as the Iowa caucuses.

The role of Core Data Scientist is one of the most interesting jobs in Silicon Valley. In recognition that people who can pass Facebook's high engineering bar and can solve problems with machine learning are a rare and incredibly valuable breed of engineer, the Facebook Engineering org breaks them off into their own job category and imbues it with special privileges that would make the right kind of engineer's mouth water. They don't work for teams per se but for the company at large. They can choose the problems they work on, and they are assessed only on the impact they drive. It's a degree of freedom that is rare at any company in Silicon Valley. They are like knights-errant, riding through kingdoms and slaying convenient dragons along the way. For a product manager, the greatest upside of core data scientists is that they have the organizational flexibility to magically appear and transform your project. The downside is the same: They can magically appear and transform your project.

Within my team, I had expressed concerns that such a trip would cost us precious time, and there was the possible chaos factor Sasha might throw into the project, but our team's engineering manager decided the trip would be a go. Word of the trip reached Samidh only after they had bought nonrefundable plane tickets. He was furious with me that I hadn't stopped it. He pointed out that there was only one Facebook employee in New Zealand; if we all wanted to cowork, he should have flown here and we should have rented an Airbnb. Samidh was preaching to the choir. I felt like I was just strapped in and along for the ride. Little did I know where the chaos machine would take me next.

When the end of December came around and it was time for us to review our First Half, 2020 road maps with the leaders of the Integrity org, I was just grateful I had survived the past six months. We showed up in a large conference room with the heads of all the large project areas within Integrity, and we walked through our plan. We explained the progress we had made so far that quarter, what we expected to have by the end of the following June, and how we expected to get there. Guy Rosen, vice-president of Integrity and my manager's manager, threw out the first question.

"That's great, but what will you have ready for Iowa?" By "Iowa" he meant the Iowa caucuses, coming on February third, the first contest in the process of picking candidates for the US presidency.

I flinched internally but responded, "The research team says there is little if any evidence, historically, of geographic targeted narrowcast misinformation, but there are lots of examples of sociographic targeted misinformation. Given our very limited resources and tight time frame, we had to triage hard." I paused for a moment and continued, "But I'll tell you what, let me go look around and find out what's possible by Iowa and I'll get back to you." He seemed satisfied with the answer and moved on to other questions.

"For Iowa" meant what was going to be possible in the Facebook war room during the week leading up to the caucuses and the day of. I reached out to the program manager for the team that was writing the Command Center software for running the war room, which in Facebook parlance was called the IPOC—"Integrated Product Operation Center"—so as to not glorify actual war rooms.

We sat down the next day and I was floored to find out that while the Command Center platform could do many different tasks relevant to monitoring an election, it had been designed for national elections like India's or the European Parliamentary election, where each population being monitored was massive—the size of a full country, not a subnational governmental unit like a mere American state. They had never run into this as a problem before. Due to schedule delays, generally speaking, the war room software would not be able to examine Iowa in isolation until *after* the caucuses were over.

This seemed like a big problem. Iowa has less than 1 percent of the population of the United States. Any targeted efforts there would be lost in the noise at the national level. While in general I agreed with the assessment of our researchers—that sociographic misinformation was more important almost every day of the year—the Iowa caucuses and probably the New Hampshire primaries were the only small-scale elections influential enough that a bad actor would geographically target misinformation campaigns. There are numerous examples of Iowa and New Hampshire reshaping presidential elections. Barack Obama, for example, had been a long-shot candidate until he came in first at the Iowa caucuses. Taken together, six of the seven presidents elected since 1976 won the Iowa caucuses or the New Hampshire primary. The Iowa caucuses have been the first contest only since 1972 and didn't earn their kingmaker status until Jimmy Carter's strategy of focusing on the caucuses and early primaries helped him win the nomination four years later.

Focus your targeted misinformation campaign on Iowa and you have a good chance of influencing all the subsequent caucuses and primaries. How was it possible that developing the ability to segment metrics by state had been prioritized so low that it wasn't going to be ready by Iowa? And to make things worse, we had just two full workdays left before everyone would leave for three weeks of vacation.

———

The level of conflict I had lived through over the previous six months was like nothing I had experienced at any point in my career, and I was now beginning to freak out about Facebook's possible impact on the election. As we entered the holidays and the holiday party season, I would sit at home, fully dressed and otherwise ready to go, and procrastinate about leaving my apartment to attend events I was invited to. It wasn't that I didn't enjoy socializing — as soon as I got there I wouldn't want to leave. But I was feeling overwhelmed enough that I would sometimes sit at home and cry while trying to build up the strength to get out the door.

The pre-party procrastination had developed at some point over the previous months. In San Francisco, everyone asks you about how work is going or what you do for a living. Every time someone asked, I couldn't escape the feeling that I was carrying a secret, or rather secrets, about what was going on inside Facebook. Secrets about dysfunctionality and unpreparedness, both organic and willful. And that the scale of the problems was so vast I believed people were going to die (in certain countries, at least), and for no reason other than higher profit margins. How could I go to a party and just shrug a reply when I knew our democracy was being poisoned, perhaps terminally so? I felt powerless to do anything that would in any way help avert looming worst-case scenarios.

On New Year's Eve, for the first time in my life, I had a full-blown panic attack. It felt like even my body was betraying me. I sat on the edge of my bed, my lungs expanding and contracting as fast as they could. It was almost as if they were always a half beat behind the oxygen my body demanded, no matter how much they raced to keep up. I was living what destroys most whistleblowers — I was alone, carrying a secret that I believed could kill people.

By the time I had exhausted myself enough to regain my footing, it was 1:30 a.m. The ball drop had come and gone. I forced myself into an Uber and went to see if there was a tail end to one of the two parties I had planned to attend. It was exactly what I had expected — funny, kind, smart people hanging out and having a good time. The nucleus of the party was a group of friends who had gone to Princeton together twenty years earlier.

Someone asked why I had missed midnight and had only just arrived. I told the truth: I had been busy having a panic attack. People were incredibly supportive, and some even shared their own experiences of feeling overwhelmed, especially during the holidays. It felt liberating to live in the truth.

The Civic Integrity team returned from winter break in full force. I was meeting with the Command Center team every couple of days as we swung into the caucuses. Now that more people were around, their engineering manager joined and explained to me the development plan for adding state-level support. It wasn't going to be ready in time for Iowa, but it would be ready by New Hampshire, a week later, or at worst Nevada, another eleven days after that. Finally, some good news.

Which states were they planning to include misinformation support for and when? They claimed they "couldn't" do all the states all the time. I put *couldn't* in quotes, because they could, they just claimed they "couldn't" get the data center budget allocated for it. It was a classic Facebook false choice. I assumed they would do some sort of rolling schedule, perhaps begin by covering the first ten primaries and then make space for new ones as they passed. I was floored when they said they were going to support only the swing states as defined by the Cook Political Report—an established scoreboard of which states will be particularly competitive for the presidential election.

"Wait, I'm confused," I said. "Why are you supporting the swing states?"

"Because they're the ones that will matter the most in the 2020 election," the program manager responded.

"Yes," I said. "But only the Democrats are having a genuine primary. Trump is the Republican candidate." I thought my point was obvious.

The program manager looked at me with a blank expression. She didn't get what I was trying to convey, and asked, "Why would that matter?"

I drew in the dotted lines that I thought were Civics 101: "Who wins the primaries will determine who we get to vote for in the presidential election. And if you only do the swing states, you'll leave out where the most Democrats live, like California and New York. If you win early primaries you look like a winner, which makes it easier to win later ones. Focusing on the swing states will functionally leave the Democratic primary process unnecessarily exposed." She seemed to be listening and considering.

I checked back in with them a day later and found out they were going to stick with their plan. I choose to believe the most generous interpretation here, which is that their team was almost certainly stretched too thin.

In my next one-on-one with Samidh, I raised this issue, and he concurred that there needed to be rolling support for the primaries. I didn't have to give him any extra context; he had worked on civic software for years. He understood how the primaries worked and the impact on a general election. "Who's responsible for getting this fixed?" I asked. "You are," he said. "You're the segmentation product manager."

Before the break I had pitched Samidh a new way of thinking about the work our team was doing. I had pointed out that narrowcast misinformation might be only one application for the software we were building. We were constructing a way to segment the United States into subpopulations to see if people were running targeted information campaigns, yet that segmentation might have many other applications. Every Integrity harm on Facebook has what are called an initiator and a target. People who generate significant genuine hate speech are not randomly allocated throughout the larger population, nor are the targets of that hate speech. One problem well known to marginalized communities is that often their content is unfairly taken down when it's not violating a platform's policies. This is called "over-enforcement." Facebook's Integrity systems used simple ways of interpreting language that struggled to detect when a population had reclaimed a slur, and usually couldn't tell when a speaker was talking *about* hate speech versus when they were *committing* it. For example, in the time since I came forward, I have talked to numerous activists around the world who have had their Facebook accounts taken down because Facebook's systems incorrectly thought the activists were harassing women, instead of *being* women raising awareness of violence against women. If you could enhance the linguistic analysis by categorizing the subpopulations the speaker and the audience belonged to, you could substantially improve the accuracy of your enforcement. And just like that, we had become a Segmentation team with a narrowcast misinformation project.

Now I was regretting that pseudo promotion. It wasn't like this was the only caucus-related work that I needed to get done. I was already flagged for being behind on writing the "playbook" the war room was supposed to use to find and deal with narrowcast misinformation. This, despite the fact that just a month before, we made clear we were only promising the ability to detect narrowcast misinformation by the end of June, not February. I didn't know what to tell them. This was a hard problem — the core Misinformation team had intentionally not worked on it for the past four years for a reason.

I was also trying to minimize the damage the New Zealand trip was going to cause. Samidh had quite reasonably demanded that there be a plan for what the team would work on while they were in New Zealand, along with a list of target deliverables. The only challenge with this goal was getting Sasha to agree to a plan in advance via a video conference from New Zealand. I worked with him and the team to hammer out a goal and deliverables based on what the Russians had been caught doing a few months before. Whatever we built needed to be able to identify behavior that matched the Russians' TTPs (tactics, techniques, and procedures). If the system could not find people who were disproportionately sending information to the same subpopulations or to improbable combinations of subpopulations, the system was not a narrowcast misinformation system.

I went into the week leading up to the caucuses holding my breath. I was slotted into the war room on Monday and Tuesday, the day before and the day of the caucuses. And it was a juggling act between trying to work in person with the teams that were in the room and trying to minimize how much Sasha pulled the team off-track from sixty-five hundred miles away. It didn't matter that Sasha had been on the video conference calls when we had nailed down the plan with the team while they were still in California. Sasha was building a science fair project that only vaguely resembled what was in the planning doc, and it would not detect future information operations that behaved like the Russians from September.

I had been helping the Civic Product manager within the Groups Integrity team prep for the caucuses, and I decided I was going to help them with their work for the rest of the day. It felt so nice to be in a low-drama environment. I thought back on an inflection point from last July where I almost called up Search Integrity and left Civic Misinformation. I would have given anything in that moment for a time machine to go back and have a do-over.

The February Iowa caucuses came and went. We did not have the tools to detect whether there had been targeted misinformation, so I can't say if it was or was not there. Another primary would be held next week, and the team would be back in the States around then too. We were going to pick ourselves up and move forward. Naturally, there was much media coverage of the Iowa caucus. After all, Pete Buttigieg, a little-known mayor from Indiana, had come out of nowhere and beaten the likes of Senator Bernie Sanders of Vermont, who finished second, with Massachusetts Senator Elizabeth Warren in third. Not to mention Joe Biden, the future president, in fourth.

Those headlines were soon eclipsed by news of a novel virus. I had begun tracking this new respiratory virus out of Wuhan, China, starting on January 21, when the *New York Times* wrote about the first case being found in the United States. A friend and I had been exchanging news clippings and journal articles that nervously discussed the few whispers that were escaping across the Great Firewall. We started a group chat of six or seven friends to discuss the emerging concern. The day after the caucuses, we decided there was so much information coming out and there were so many of our friends who wanted to be informed about it that I set up a closed Facebook Group to share information and data, and to begin to build a network of support. That group and network would prove invaluable in my future.

I had glimmers even then that COVID-19 would be extremely disruptive, but I could not have imagined what was about to come. I had no idea that a respiratory virus so far from Menlo Park would eventually lead me to share my secrets about Facebook with the world.

CHAPTER 10

Escape from San Francisco

Both individuals and societies tell themselves stories to simplify and make sense of the messy chaos of reality. It is naive to think that it is possible to live without the protective bubbles these stories create. But sometimes the stories can become terribly limiting and trap us, and prevent both individuals and whole societies from moving on into another kind of future.

—Adam Curtis

Winter kills people. It's one of those things Midwesterners will tell you plainly. Every blizzard brings news reports of people who made foolish decisions to roll the dice on a snowy road. You can have whatever opinions you want to have about the cold, but the cold has opinions of its own.

In other words, when I found myself on I-80 driving through the Sierra Nevada during a blinding snowstorm in a Prius station wagon, I had no defense. I knew better. But that's where I was. I could barely see the cars ahead of me. The next thing I knew, my Prius was spinning out of control. In my rush to escape San Francisco, I made time to stop along the way and put snow chains on my tires, but as my car spun and I wondered how this moment was going to end, the chains weren't much help. Or maybe they were, and it was just too icy. After two full revolutions, my Prius stopped spinning. It was a brief, terrifying moment that made me all the more cautious about the long ride before me.

I was racing east. I wanted to beat what was likely to be the official closing of I-80. A day or two earlier, I had gotten a credible tip that San

Francisco was going to announce a COVID-19 lockdown. There had only been two major lockdowns at that point—Wuhan, China, and in northern Italy. Both had included bans on nonessential travel. I didn't want to be trapped for who knows how long in my pseudo studio apartment, with its shared kitchen and COVID-19-skeptical co-tenants. I decided I was going to join my parents in Iowa.

The tip about the lockdown had come secondhand from the chief technical officer of one of the largest HMOs in the Bay Area. The information had found its way to me because of the closed COVID-19 Facebook group that I had started the day after the Iowa caucuses. Once that page was up, it grew rapidly from a hundred people to more than three thousand members. Even at the start of February, when the group first formed, the information coming in about the novel virus was ominous.

On top of my COVID-19 angst, following the Iowa caucus things weren't great for me at Facebook. Two days after my shift in the caucus war room had passed, Samidh called me to a conference room for my second-half performance review, which effectively covered my first six months at Facebook. That there was a human resources person sitting with him was a sign that my review wasn't exactly going to be glowing. And it wasn't. I was held responsible for the fact that my team didn't have a working prototype for our narrowcast misinformation detector by the end of the 2020 election lockdown. That we didn't have a narrowcast misinformation detection playbook for the caucuses. The team just wasn't productive while working with Sasha in New Zealand—and that was apparently my bad. I had pleaded that the team not make that trip for that very reason, but that didn't matter in this review.

What I found most frustrating was that I was being dinged for having "conflict" with the Command Center team because I had multiple conversations with them about covering the Democratic primary states instead of the swing states. The only reason I had persisted in this was because Samidh had told me it was my responsibility to get it fixed. I calmly tried to present context for all of this, but quickly resigned myself to the fact that there was no point. I had questioned the senior program manager whose domain the Command Center team fell under, and she had complained. She outranked me, so that was that. During the review, Samidh informed me that they were going to follow my advice and find a new engineering manager with algorithmic product experience for the narrowcasting team, but to show that they weren't favoring me above the engineering manager, we both were being taken off it. I would have six weeks to pitch a new project that was "worth" staffing with engineers.

Fortunately, I had in mind what I believed was indeed a worthy project and had already begun doing some due diligence work on it. What had made Sasha interested in narrowcasting was a meeting he and his collaborator had with me to discuss their project around influence modeling. Often theoretical projects have trouble finding ways into production systems — you need a good product translator who can find the right place to slot in your new technology in a productive way. I thought their first application should be to find narrowcast misinformation. They were thrilled that I wanted to be their product manager.

Despite the tensions over New Zealand, Sasha and I had found a common purpose with our interest and work on influence modeling. While the Segmentation team had primarily focused its narrowcasting efforts on situations where a bad actor would send targeted misinformation to a subpopulation of the United States, another equally likely scenario was even more troubling: bad actors targeting high-influence people within a subcommunity or the nation at large. Think police officers and environmental activists. Despite being overruled by Mark Zuckerberg, Facebook's researchers had been making the case that political leaders should be held to the same content moderation standards as everyday users because, of course, average people are more likely to believe information that comes from leaders they already trust.

When it came to narrowcast misinformation targeted at a subcommunity of the United States, a Russian influence operation might involve something like sharing a series of posts to a collection of Facebook groups that all focused on the same population. They might then follow up and have a network of accounts all comment on those same posts in order to create the perception that people within that community agreed with the sentiment provided. They might target ads at people who are members of the same group. They might use "virality hacking" to build Pages (or Groups) that appealed to their target population by identifying posts that had gone viral before, then reposting the same link or image on the assumption it will get more distribution than the average piece of new content they might create. Once the Page or Group gets popular enough, they start to sneak in disinformation so it's distributed to the audience alongside the content they followed the page or joined the group to receive.

Narrowcast misinformation targeted at leaders likely would take different forms. The influence operation might directly message people they believed to have influence, befriend them, and send them links to information that was misleading. They might friend them in the hope the target accepts the friend request, so the attacker could share information into their feed. They might aim the same types of activities at the friends of their target, with the hope of working toward friending the target once they had enough mutual friends to lend credibility. They might target ads at specific lists of people—a concern Facebook was aware of, though their defense against it was to require such custom audiences to have at least one hundred people in them. Not exactly effective if the attacker adds ninety-nine people known to have been inactive on Facebook for a long time. If we could identify who had influence on Facebook (or out in the world), we might be able to identify those who were trying to narrowcast to them.

Our influence modeling project was implicitly a critique of Facebook's ideological belief that everyone was equal on the platform. That they had created a flat world where all voices were the same, and hypothetically anyone could reach a million or more people. Facebook's obsession with having a flat office and flat organizational hierarchy was in some ways a reflection of their idealized vision of the product they thought they had built. Regardless of the intensity of the leadership's belief in this dream, Facebook Blue was not flat at all. It was basic common sense that some users had more influence than others, and we believed there were many ways to detect and quantify that. On a crude level, you could count how many views an account attracted. Facebook had used measures like that before, for example for the Cross-Check program, which effectively exempted certain power-users from content-moderation policies.

We wanted to get more sophisticated, more granular, and look at things like if a person was being mentioned in political posts or being searched for in the same context as others who were already identified as civic actors (people with outsized influence in politics or civil society). Were they friends with known civic actors? Did known civic actors send them messages, etc.? Even things like "Do others message you more than you message others?" had implications for influence. This would imply that people want to talk to you more than you want to talk to them.

Sasha had been at Facebook long enough to understand the value that poaching a product manager would have for a core data scientist who wanted to get a project like influence modeling off the ground. He was spinning too many plates to advocate for engineers to staff this pet project, which he had begun with a research scientist named Theo. If he could get me on board, recruiting more support to the project would become my responsibility. And so he filled me in and asked if I was interested. Recruiting me would increase the chances he would actually be able to build a team around influence modeling. I shared all of this with Samidh, and he informed me that in six weeks I would need to present influence modeling to Guy Rosen, head of the Integrity Organization. It would basically mean making a pitch to save my job.

One of the things I soon discovered was that within Facebook there was no archive of examples of what was expected in a "good" project pitch; for that matter, there really wasn't a company-wide archive of documents of any kind. Inside Google and Pinterest (the latter largely founded in the beginning by ex-Googlers), there was a search engine for such documents; and it looked remarkably similar to the Google available in the outside world. You could put in a search term, and ten blue links would show up that would present potential templates or existing resource materials. Inside Facebook, there was a search engine like that, but it covered only a *sliver* of the documentation that existed inside the company.

The primary form of institutional knowledge management was a product called Workplace. Workplace in many ways paralleled what Facebook looked like to the public. You could follow people or groups, or in the parlance of the veteran Facebook employees who still often referred to groups by their original name, "tribes." Workplace search extended across publicly shared content and the specific closed groups you were already a part of.

When I searched for examples of project pitches, nothing came back. The core difference between accessing information with a traditional search UI with titles and snippets versus a stream of content is it's much slower to look through twenty, thirty, forty, or more documents when you must slowly scroll past each of them in a feed. When I asked others in Civic Integrity, crickets. Same with the internal group chat with the twenty or so Olin alumni who now worked at Facebook.

Every organization has its own invisible expectations. If you want to ensure that the widest spectrum of people succeed in your organization, you want to accumulate examples of great work and best practices that others can mirror. At Facebook, despite my hunting around, I found only one example of a project pitch. It came from my manager, and only after I explained my futile efforts to find one.

Meanwhile, I began trying to assess what tools, teams, and efforts already existed within Facebook that did work related to identifying high-influence users. I needed to understand the business cases where Facebook was already providing extra protection or was exercising more caution toward some users over others.

———————

My communication with my parents had picked up because of the COVID-19 Facebook Group I'd started. My cofounder and I had invited maybe a hundred of our closest friends and added my parents. For thirty years my father's research had focused on viruses, and my mother, when she was still in academia, had been a cell biologist. Initially our focus was simply on the questions of how dangerous this virus might be and whether COVID-19 had escaped containment in the United States. Did we have evidence of community transmission? People would submit papers attempting to estimate how infectious the disease was to help us gauge how fast it would spread, or papers trying to tease apart the risk factors for serious illness. Both of my parents stepped up, answering questions people posed about the research or about pandemics and disease in general.

One of the papers that most alarmed me described the "physics" of the disease, also known as the kinetics. How long did someone incubate the disease before they showed symptoms? Once they had symptoms, how soon and how often would they be admitted to the hospital? Once they were in the hospital, how soon and how frequently would they deteriorate enough that they needed the ICU? Once they were in the ICU, how soon and how often would they die?

If COVID broke containment, the question was not whether the virus was going to blow up, the question was when. Even early on it was known that every person infected with COVID infected at least one more person, the definition of a disease that will continue to spread on its own. The biggest critical factor in determining whether containment had been breached was understanding how many cases were already in the United States. Ever the data scientist, I coped with my anxiety over the growing threat by using the new "physics" of COVID-19 to predict how many cases were hidden beyond the view of the limited testing available in February 2020.

Humans are incredibly bad at internalizing how fast exponential growth, like a virus breakout, unfolds. If you really want to trash people's ability to intuit what the future will be like, just throw some time lag into the mix. When a quantity doubles consistently as time goes by, it is a form of exponential growth. If you double every day, at the end of the first week you have 128 times what you had at the start. But at the end of the second week you have 16,384 times as much. It's almost impossible for people to imagine 16,000 is the number that will come after 128. Now imagine you were unable to see the state of the world as it is today — you could only see the world as it was a week ago. You thought you were dealing with a 128x increase, but you really had an explosive 16,000x increase on your hands.

Best-case scenario, people were not getting tested for COVID until they got ill enough that they ended up in the hospital. This would mean there was roughly a twelve-day lag between infection and testing. Add in another three- to seven-day lag as those samples traveled to Atlanta to the only testing facility and were processed, and every reported case was really a signal from the past of an epidemic that had already run away. To make it even worse, in a world where testing was precious, hospitals were often waiting until patients were so far gone that they were transferred to the ICU or even died before testing them. Now the signal was an echo from an even more distant and misleading past.

With my spreadsheet analyzing the kinetics data, I could make guesses about how many more cases were hiding in that lagged exponential growth. The results were shocking enough that I messaged my parents and brother and asked if they would be willing to peer review my spreadsheet the next day. We all dialed into a conference call the following afternoon and I walked through the logic of my spreadsheet. For every case that escalated to the point of ending up in the ICU, there were 100 hidden cases of fresh infections that were working their way down the pipeline.

Yes, there had been only about ten positive cases announced so far, but realistically it seemed unlikely that even 10 percent of the cases that had advanced to that point had been caught. It seemed likely there were hundreds if not over a thousand more cases in the wild when all factors were considered. COVID had escaped. People just hadn't realized it yet. My parents and I agreed that afternoon that I would come and lock down with them in Iowa. I bought my Prius a few days later and started planning how and when I was going to pack up my things.

I didn't know it yet, but this moment, sitting there with my parents walking them through my vision of the future, would set me on a path to being a whistleblower. Armed with math at my back, I felt for the first time that I needed to convince the people I cared about that the world was about to change rapidly. What came next gave me the evidence of and confidence boost in my abilities I needed to endure at Facebook and later blow the whistle.

I felt compelled to convince my friends because the COVID-19 group seemed to be bifurcating into two camps: those who were trying to convince people something historic was coming, and those who thought those people were catastrophizing. One of the early joiners who I felt had a calm head but who was certain that the world was in for a shock was a friend, Amelia. She had begun to talk about shorting the market. In theory, I knew what she meant. If you believe the price of a stock is going to fall, and fall significantly, you can bet on that, even if you don't own any shares. You simply borrow them often at very little cost, and when the price falls, you can return what you've borrowed and pocket the money generated by the price gap. The risk is that if you're wrong, you could be on the hook to pay for the full value of everything you borrowed or more. It's called "shorting" the stock.

Amelia was doing something slightly more sophisticated. The other way you can "short" a stock is to buy something that acts almost like insurance on it — it gives you the right to sell a stock at a given price. It is called a put. Amelia had mentioned almost immediately after the group formed that she had bought a bunch of so-called puts that were "out of the money" — their trigger price point was substantially lower than their actual price. Stocks rarely drop by 10 or 20 percent over the course of even three months, and as a result you could buy these "out of the money" puts for a penny or five cents. If prices crashed by 10 percent, suddenly those pennies might be worth ten or twenty dollars.

I didn't pay much attention to her at first, but as the market continued to go up through the first half of February, I sprang into action. We were now a couple of days after the spreadsheet that had set me in motion. I called up Amelia and asked for a refresher on finance. I had been so hostile to finance when I was in grad school that I had just signed my name on my second semester final exam and walked out. I knew if I got the equivalent of an F it wouldn't impact me at all because Harvard Business School didn't report grades. I had other things I wanted to do with my day.

Now I desperately needed to refresh my understanding of how puts worked. On Monday I sold all the stock I had in my investment accounts and waited until Tuesday, when the money would be available. I woke up at six and called Schwab before the market opened to make sure I understood how to buy what I wanted to buy, and then I put all my chips on the table. The next day the market went up, but the value of my puts also went up. They doubled or tripled. I sat there staring at my computer in confusion. This wasn't how it was supposed to work. Watching the market rise while insurance against the market falling *also* rose was a confirmation that I was on the leading edge of something. The people who were watching COVID-19 spread felt confident they knew what the future held, but for the average investor, institutional or retail, we were still living in a bull market. That day turned out to be the peak of the market. The S&P closed with a record high on February 19, 2020, at 3,386. Three weeks later it had fallen 26 percent, to 2,480.

In the end, despite the value of my puts increasing by more than twelve or thirteen times, I didn't make that much money off it by the time I sold because I didn't actually understand finance and I didn't know when to sell. In every previous pandemic, the stock market continued to sink until the case count peaked. I held tight to that theory and didn't sell, even after the market turned around. I didn't know or understand what it meant when the United States and other developed countries began to print so much new money that they bought their way out of the depression that likely would have resulted if they hadn't stepped in. I held on too long and only doubled my money, not getting the kind of life-changing return the one friend I coached made using the same strategy. He was smart. He recognized that he didn't know what he was doing and sold right away, as soon as the value rocketed up. He bought a six-thousand-square-foot house in Berkeley with his windfall.

That was still in the future. In the last week of February 2020, I couldn't know that my successful bet on the market would be one of the major factors that would help me decide to blow the whistle the following year. I had seen something that few others had seen. I made a call. And I was right. I had spent seven months at Facebook being gaslit and told over and over that I was being dramatic or ungrounded in my observations, and now I had proof that I could take in data and make real predictions about the future. I feel certain today that if I hadn't had this experience, I never would have blown the whistle on Facebook. I had found my confidence in myself again. I can't stress this enough: anytime you find yourself in a situation where you think you are being gaslit, when you think people are actively making you doubt reality, the most important thing you can do is find something that allows you to test and ground your perception of what is real.

As I prepared to make my trip to Iowa, I talked to my manager about COVID-19. I said that I didn't understand why Facebook was making no motions to respond to it. The Bay Area was one of the only places in the country with confirmed COVID-19 cases. What was going to be the plan for working from home? How long could the buses run? He had no answers for me, but the next week at our weekly Civic Integrity product team meeting, I won the award for upholding the value of "caring" for my coworkers.

This period was particularly stressful for me because it felt like we were in the quiet before the storm. I still had many close friends in the COVID-19 group saying we were overreacting. For the first time in my life I felt I really understood why the Greek myth of Cassandra was a tragedy in the literary sense. She was given the gift of prophecy by Apollo but was also cursed because no one would believe her. I felt a constant pain in my chest for the first ten days of March as I lived in that tension. This in turn certainly influenced how much documentary evidence I copied from Facebook when I became a whistleblower. It wasn't just that after having my concerns second-guessed so often by my coworkers I didn't think anyone would believe me. It was also that I had witnessed how steadfastly people will resist information that's uncomfortable, even if it comes from a trusted friend. I needed to make the case for the problems of social media so concrete and robust that it couldn't be doubted.

When the text message came in that the lockdown was imminent, it felt like a relief. My job was done. I had tried to warn people, and now the storm was here. It wasn't my problem anymore. I frantically packed the last of the things I would take with me into and onto my car and set off for Iowa.

I ended up hitting the road around three in the morning. I hadn't expected the lockdown announcement to come that early or I'd have packed more in advance. I had to leave that night because I knew I was racing a monster of a winter storm. If I couldn't get beyond Donner Pass before they shut the road down, I'd be sent back to Sacramento to wait out the storm, potentially for days. All the same, by the time I crossed the Bay Bridge I realized I was tired enough that I couldn't safely drive. I took a nap in my car in a rest area in Hercules for ninety minutes and set out again in the morning.

By the time I reached the mountains, the snow was falling in force. My childhood in Iowa had taught me to respect the weather and the dangers of the road. Snow kills people every year. It's not just car crashes. It's people who try to walk for help after their cars get stuck and freeze to death. It's people who do the safe thing and stay in their cars to wait for help, but still lose fingers and toes to the cold. My childhood friend Tina was killed on a slick road. When I hit the snowstorm and couldn't see the cars in front of me, I knew the sensible thing was to turn back, but the fear of being stuck in California if I didn't get out now seemed like a bigger risk. All of this flashed through my mind when my Prius started to spin. Yet when it stopped, I was still pointed east, toward Iowa. I continued heading toward Nevada, a little slower this time.

I pulled into the parking garage for the Nugget casino in Reno around 1:30 p.m. and took a nap. I had bought a Prius V, the now-discontinued station-wagon version of a Prius, because there was an online culture about converting them into minicampers. I vaguely aspired to creating a Vanlife van one day, and starting small, with something that could easily be converted into a vehicle that could be slept in, seemed like the right first step. I had spent an evening earlier in the week building a sleeping platform in a friend's garage that gave me a pocket that I could stretch out in to sleep.

Waking up in my car felt magical. I had my own warm little self-sufficiency pod. The past month I'd felt like the world was out of control, and waking up in the parking garage in the process of executing my plan was the first moment I felt grounded. I had made a bet that this would be a workable option for traveling cross-country, and my bet had paid off. I had a heated blanket I could plug into the cigarette lighter that drew just a trickle off the mighty Prius battery. Combined with a good sleeping bag, this made me a toasty little voyager. And I felt vindicated by my rushed escape from San Francisco. Just ninety minutes after I crossed the pass, the California Highway Patrol had closed the road due to spinouts.

I pulled out my phone and informed my parents I was awake, okay, and driving onward. Before I set out for Iowa, my dad's only ask was that I turn on my GPS location sharing so he could watch me as I traversed the country. I realized early on in my drive that he was actively watching. If I spent too much time at a gas station or stopped to rest for a little while by the side of the road, I'd see an instant message from my father asking if I was all right.

Time on road trips tends to blur, and I drove eastward for another day, my mind wandering as I crossed the expanse that is Utah through the slowly waxing hours of light. I stopped to take photos at the Bonneville Salt Flats and marveled at how heartbreakingly beautiful the desert was at twilight. No matter what humanity would face in the next few years, the

Earth was boundlessly older than we were, and would remain long after we were gone. I pressed onward and crossed into Wyoming. Late in the night I passed through Rawlings, and my mind went back to the last time I had made this drive to Iowa, that time with my brother, before starting graduate school.

Out of the blue on that drive eleven years prior, the car began to buck. We had a flat tire. Standing by the side of the highway under an unimaginably big sky, I surveyed the empty rolling hills and wondered what Peter and I were going to do. Neither of us had cell phone reception nor knew how to change a flat (at least not without help from YouTube). Cars rarely passed in either direction. Then out of nowhere, not more than fifteen or twenty minutes later, a car stopped and the kind couple inside offered to help change the flat. They gave me probably the most important advice I received that trip—I needed to press forward toward Laramie. Yes, Rawlings was only twenty-five miles behind us and Laramie seventy-five ahead. And yes, our spare was only rated to fifty miles at fifty miles per hour. But they were the owners of the only store in Rawlings that sold tires, and they were on a day trip to Laramie. Our only chance of getting a new tire lay in front of us, not behind us. They were clear. We needed to risk it or accept the prospect of being stuck until they returned home.

There are these little moments in your life where you either come to believe in or question the kindness of other people, and that moment on the rolling plains of Wyoming was one of the times in my life that made me believe in the importance of people caring for each other. These folks didn't have to stop for us. This was their day off, but they knew no one else was around to help. They could act to keep us safe, so they did. It was moments like these I recalled when I was trying to decide whether to blow the whistle.

As I pushed ahead in the darkness, I hoped I wouldn't have to test the willingness of my guardian angel, or the people of Wyoming, or Nebraska, or Iowa for that matter, to intervene again. I still had a long way to go.

I pulled into my parents' driveway at 2:30 a.m. on March 18. Driving alone and through a winter storm, it had taken me ninety-three hours to cover 1,912 miles. At the side of the driveway was a sign. When it had become apparent that COVID-19 was going to be an issue that would test and quite possibly tear America apart, my mother had pulled it out of storage and stuck it in the front lawn.

You are loved.
Your life matters.
We are in this together.

I got out of the car exhausted, took a picture of the sign with my phone, and posted it to my COVID-19 Facebook Group. I hoped it would serve as a reminder to my friends that whatever was before us, it was not the end. This was just the beginning, and indeed we were in this together.

I unloaded the Prius, quietly crept up the stairs to the guest room, and collapsed into bed. The next morning when I came downstairs, my parents were sitting in their sunroom, a two-story space they had added soon after I left for college. It was cold and wintry on the other side of the glass, but inside, the room was warm and filled with sunshine. My mom and dad got up from their chairs, walked over, wrapped their arms around me, and said, "Welcome home!"

CHAPTER 11

A Decentralized Threat

The most beautiful thing we can experience is the mysterious.

—Albert Einstein

According to the six-week deadline Samidh had given me, I was scheduled to pitch the influence modeling project to Guy Rosen, head of the Integrity Organization, on March 16, 2020. That, of course, was also a pitch to save my job with Facebook. Yet on that day I was behind the wheel of my Prius, en route to my parents in Iowa. My decision to leave the Bay Area when I did turned out to be a good one, as on that same day Facebook announced it would close its offices in accordance with a state-mandated lockdown. My pitch was postponed. The lockdown gave me, Sasha, and Theo more time to prepare our case.

As with the rest of the world, COVID-19 and the lockdowns disrupted much within Facebook. Since it was a software company it was easier to work remotely, but the company wasn't immune to the impacts on its business. Some staff lost their childcare. Employees who lived in shared San Francisco homes had to reshape their bedrooms into "everything" rooms, where they now worked and slept, for who knew how long. Like many companies, Facebook suddenly confronted reduced head count as employees went on temporary leave while they figured out how to cope. Meanwhile, the company also had to deal with additional, once-unimaginable challenges. Many new projects born of the COVID-19 crisis immediately took precedence on the high-priority list.

Sasha was swiftly and completely occupied on COVID-19-related projects, working hard to produce privacy-sensitive mobility datasets to help governments ascertain how well communities were adhering to the lockdowns, staying at home and reducing their travel. His collaborator, Theo, was working on Facebook Connectivity, which is to say, trying to expand the internet (and more specifically Facebook) to more places and people around the world. Free Basics fell under the Facebook Connectivity umbrella, so Sasha and Theo managed to squeeze in time to compute experimental influence modeling datasets and then hand them over to me to validate and critique.

Back in Iowa, I was available all day to focus exclusively on influence modeling. At home with my parents, a routine quickly developed. I would wake in the morning, and because I was quarantining on the second floor, pick up the breakfast my mom would leave for me on the stairs. I'd browse updates on Workplace as I ate and then dive into my work. I realized we might really be onto something big while I was examining the influence levels of people who initiate friend requests and those who are the targets of them. When I transposed those findings into a graph, the story was compelling. The graph was one of the most beautiful I've ever generated.

I looked at people with different influence scores, and at whom they typically friended. On average, people initiated friend requests with people who had approximately equal influence. People with high-influence scores sent friend requests to similarly influential people, and people with low-influence scores sent requests on average to low-influence people. Makes perfect sense—birds of a feather, after all. The part that made my pulse quicken was that on average the person who *received* the friend request had slightly more influence than the person who sent it. We defined "influence" as the ability to attract a response from others, and our model, despite not yet using friending as an input, had successfully found a pattern of influence impacting friending. This was a giant thumbs-up that we had found a real opportunity.

Samidh was sufficiently impressed that he said if we could find a single in-house team who wanted to work with us, we could get a charter to move ahead. He suggested we reach out to the Civic Data team and their Civic Actors project. Civic Data was responsible for monitoring and analyzing feeds from third-party vendors that aggregated data on such civic topics as the location of polling places, who the current elected representatives and candidates were, and lists of journalists or activists, among others. Facebook wanted to know about Civic Actors in order to give them extra protections.

Civic Actor status was used in many ways at Facebook and it determined the kind of treatment you might receive. With Civic Actor status, you might get priority reviews when you complained about harassment or impersonation, or you might be required to activate extra safety precautions like two-factor authentication (confirming with a text message or a one-time password-authentication app that you really were the person logging in). The highest tier of special treatment was Cross-Check (also known as XCheck). Once I'd become acquainted with Cross-Check, I remembered a moment from one of my first interactions with Samidh in a new light. He had complained that if you spent enough money on Facebook advertising, say $100,000 monthly, you automatically qualified to have your content protected. He viewed that as a direct refutation of the company's public mantra: "All People on Facebook Are Treated the Same." I didn't really understand what he was talking about at the time, but now it made sense.

Facebook publicly described the Cross-Check program as a "second check," and touted it as another level of safety. If you had Cross-Check status and Facebook's computers deemed your content offensive or otherwise unacceptable, an actual human would review that content to determine if it should be taken down. The automated content reviews had a false positive rate of about 10 percent. Facebook had launched the Cross-Check program with its human reviewers because those automated false positives generated a tremendous amount of unnecessary false "fire alarms" for the communications and policy teams when, say, celebrities or politicians had innocuous posts blocked by accident. Every innocent post that was taken down made obvious the lie that Facebook's artificial intelligence was adequate at distinguishing between good and bad content.

What was perhaps even more shocking was that the 10 percent false-positive rate occurred under the best possible conditions. Every AI classifier for winnowing good from bad content has to make a trade-off between how much "bad" stuff they can find versus how often they misjudge content and consider it "bad" when it is in fact "good"—in other words, generate a false positive. Let's imagine there are a thousand posts under consideration. Ten of those posts are some form of "bad." If you discovered five of the ten posts, your "recall rate" would be 50 percent. Every classifier can be adjusted along something called a "precision-versus-recall curve." On one end of the curve there is very low recall and very high precision—the program misses finding most of the bad stuff, but when it singles something out, it's almost always correct. On the opposite end, the machine is able to find all or almost all of the bad content, but might be wrong half the time or more when it points the finger.

Facebook tried to tune all of their classifiers so that they would be wrong only 5 percent of the time. This regularly translated to 10 percent false-positive rates once the chaos energy of real users interacted with the classifier's theoretical performance. Minimizing false positives is good—every time legitimate content is taken down, that person's right to express themselves is violated. Unfortunately, in the case of hate speech, getting only a 5 percent false-positive rate came at the cost of missing 95 percent of all hate speech on the system. This was one of the more shocking revelations in my disclosures, because for years Facebook had promoted a statistic in their "Transparency Center" implying that they were actually taking *down* 97 percent of hate speech.

Facebook's "second check" largely functioned as a blank check for those in the program. The reality was that Facebook staffed the Cross-Check so thinly that it would take two to three days for a post in the Cross-Check queues to get reviewed, and by that time the content in question would have already reached 75 percent of the people who were ever going to see it. If they had wanted to make the program at least mildly effective, they would have held the content out of circulation until it could be reviewed. For a large fraction of Facebook's safety systems, if you had Cross-Check status, Facebook just auto-approved your content because you were now on a "whitelist" of good actors. Doing a human review as a double check cost money, and the safety personnel who threw up their arms and just let whitelisted VIPs walk freely past their metaphorical gates did so not with nefarious motives, but because they couldn't obtain a budget to hire or maintain a parallel staff to review the content of the privileged.

The Civic Data team encouraged us to confer with the Cross-Check team to see if they were interested in expanding coverage using influence modeling. Less than a week later, we had both teams on board. They saw an opportunity to identify people who indeed had substantial influence but were not yet considered Civic Actors. They were particularly interested in thought leaders or activists who lacked a formal elected role, and thus a way to show up in the datasets the teams had already ingested. By the middle of June, we had been approved. Both the Cross-Check and the Civic Data teams were committed, and the Civic Data team was willing to contribute an engineer half-time to the project. We were ready to push ahead with influence modeling.

Almost immediately after we officially kicked off the influence-modeling project we ran into a huge speed bump. Sasha left the company. He had spent seven years at Facebook, and the COVID-19 crunch was the last straw. He had been running at full speed for months, and he informed us he had put in his three-week notice. He was apologetic, but said he needed to look after himself.

Sasha's feeling that he had no choice but to leave Facebook in order to protect his emotional or physical health was not unusual among employees at the company. There were countless good people at Facebook who felt they had no choice except to bail. In the process of doing their jobs, they would learn the truth of what was going on at Facebook: that the company knew its products were having a toxic and at times dire impact on society. The employee would then have to pick one of three options.

Option 1: Ignore the truth and the consequences thereof. Switch jobs inside the company. Write a note documenting what you'd found and tell yourself it's someone else's problem now. Give yourself a pass because you raised the issue to your product or engineering manager and they said it wasn't a priority. Not every problem has to be your problem.

Option 2: Quit, and know that one fewer good person, or at least one fewer person who knew the truth, was working at Facebook. Live knowing the consequences you discovered were continuing, invisible to the public.

Option 3: Do your best to solve the problem, all the while knowing there isn't enough institutional investment or will to actually fix it.

Sasha took the third path for several years. He had gone through cycles of burnout when he would work himself to the bone in a crisis, then patch himself together to try again. The crush of the additional COVID-19 work on top of everything else was just too much. So he reverted to Option 2 and departed. When I found myself at my own crossroads a few short months later, I chose a fourth option: Get the public involved. No problem is solved within the frame of reference that created it. If only ineffective solutions existed inside the company, maybe we needed the public to come save Facebook.

With Sasha gone, those who remained to work on influence modeling were Theo; Nora, the Civic Data engineer; and me. Sasha had years of experience working with the broad set of data sources available within Facebook's sprawling and often disconnected data sources. His expertise was a large part of what enabled us to get the project approved. Also, there was the fact that the whole thing had been his idea. Now I was effectively the data scientist on the project. I would wake up each day at eight, be working by nine, and go full-tilt most evenings until at least ten. This was my last chance at Facebook, and I was going to give it my all.

It started out promisingly. I had taught myself how to use a wide set of data science tools while we were validating some of our early models and experiments around calculating influence. Now I was churning out new features we could use to try to tease out who had civic influence. The first few features were smooth sailing. I pulled data on who was mentioned in posts that addressed civic topics or who was mentioned in posts written by known Civic Actors. The model produced expected results. We looked at co-occurring behaviors. For example, did someone search for a known Civic Actor or maybe even multiple known Civic Actors? If they also searched for someone who wasn't yet a known Civic Actor, that increased the chance that that unknown person might also be a Civic Actor.

With each additional feature, the model kept giving us more relevant results. It seemed it might live up to the promise we had initially pitched — that we would find people outside traditional lanes of recognition who still wielded significant civic influence. We were finding people like Suzan Shown Harjo, a Native American activist who had been a congressional liaison for Indian affairs and later served as the president of the National Council of American Indians. In 2014 she won the Presidential Medal of Freedom, the country's highest civilian honor, in part for her accomplishments helping Native people recover more than one million acres of tribal lands. And yet she was not a Civic Actor in Facebook's eyes. As I spot-checked who was appearing in our top twenty or thirty new Civic Actor candidates, I began to notice that an outsized fraction of them were people of color. Leaders of minority communities, of course, encountered more roadblocks to formal positions of influence, but that hasn't stopped people like Harjo from wielding civic influence and making the world a better place in the process. Social media platforms like Facebook were likely a critical communication tool for them, and I thought Facebook should try just as hard to keep them as safe as people in elected offices. The great irony of Mark Zuckerberg's political-speech policy was that it might protect the local dogcatcher if that was an elected position, but it didn't protect Harjo.

The progress we were making made it easy to work the seventy- or eighty-hour weeks I was putting in — until we ran into a wall. It seemed logical that we should be able to extend the results we found from looking at people who co-searched for Civic Actors by looking at people who co-viewed Civic Actors' profiles. After all, viewing profiles is often a consequence of searching (you click on a profile after you search). But instead of improving our results, it just broke the model. We went from producing output data that when visualized showed a clear distinction between people with civic influence versus people without it, to a giant tangle. By taking the next logical step in our research, we had somehow added so much noise to our model that we could no longer see meaning in our output. It was just randomness. Something was very wrong.

I took a step back and started looking at the raw data of how people viewed profiles. One reason our model could be breaking was if there were accounts that were viewing too many profiles, without a pattern to how they viewed them. There were definitely outliers. We found an account that viewed not quite 300,000 profiles in a week. That superscraper account was clearly not a human being. You would need to view 29.76 profiles every minute, twenty-four hours a day, every day, to view 300,000 profiles in a single week.

What was clearly breaking my model was scraping, or crawling — the process of using a computer program to download information from a website. So I tried again, this time excluding anyone who looked at more than three hundred profiles a week. It still looked completely random. So I dug deeper. I began thinking about how, if I were trying to be sneaky, I would hide a scraper network. I was going to figure out what was going on — it was either that or give up on the model.

You might be asking yourself: Why should anyone care about networks of scrapers? What's the big deal? So some people (or programs) look at some profiles, who cares?

On the most basic level, allowing large-scale data to be captured via scraping can have multibillion-dollar consequences. In the case of Cambridge Analytica, the FTC fined Facebook $5 billion in July 2019 for misleading the public about how their data could be accessed via the developer application programming interfaces (APIs) the company provided. *Fast Company* described how Cambridge Analytica's system recruited 270,000 users to complete small surveys in exchange for being paid a dollar or two, then took advantage of a then-commonly-used feature in the Facebook API. Without users realizing it, Cambridge Analytica harvested not only the Facebook data about the survey taker, they also downloaded similar data about that person's friends.

Just by fanning out from those 270,000 people to their friends, Cambridge Analytica was able to download data from 87 million profiles. There are widely divergent reports regarding how much impact Cambridge Analytica had on the 2016 election. While many point to Steve Bannon's involvement with Cambridge Analytica and their affiliation with political campaigns like those of Ted Cruz and Donald Trump, those same campaigns publicly questioned the value of the kinds of psychological profiles the consultancy provided. If the FTC knew Facebook was doing little to stop widespread unapproved access to users' data, would they impose an even bigger fine this time?

The real danger in my mind of these distributed networks was about how the networks themselves could be used, in contrast to worrying about how the data the scrapers harvested could hypothetically be used. I don't have documents to substantiate this, but after I came forward as the whistleblower, a Facebook employee reached out and told me about allegedly heated discussions inside the company about the role of automated accounts influencing the algorithmic models Facebook used to distribute content to people around the world. If you have a large enough automated network that can effectively pretend to be a human, those "people" can intentionally "Like," follow a link from, or reshare a post, creating an artificial perception that the piece of content is high quality. If you wanted to get fancy, you might make your fake people behave in ways similar to a target population you were interested in, such as police officers, so that you could trick the algorithm into believing police officers like certain types of ideas, further accelerating the distribution of your ideas into the actual public-safety community.

What if your network for acquiring data wasn't just simulated people? What if they were real people, with a program on their computers or phones that rode along and hid the network owner's intended behavior among the organic behavior of the humans for disguise? Open Library's project to figure out whether various books on Google Books were viewable was executed this way. They had built a plug-in that turned average people into the nodes in the scraper network.

While that might not seem too dangerous if the network was just downloading Facebook profiles, things might go off the rails if the plug-in started sending out misinformation posts or comments pushing a narrative. Imagine if the day after the 2020 election (or any other election), ten thousand people around the United States all started posting about seeing people destroying ballots while tens or hundreds of automated people jumped on their posts and left comments saying that they were there too or they saw the same thing. Finding such a distributed network might be your first, and perhaps only, warning sign of a much larger danger.

———————

The only way to catch a scraper is to think like a scraper. I began to work my way through how I would design my network to hide what I was up to.

Let's imagine you have a target list of profiles from which you care about downloading data. Maybe they're profiles your subscribers want monitored because they belong to influencers, and your subscribers want to see what the influencers say about the subscribers' products. Or maybe you're just making a big database of all the users on Facebook and you're going to download as many profiles as possible so you can sell the aggregate data to marketers or politicians to mine. Maybe you're a national security agency and you want your own database of Americans who have any interest in your nation.

If you want to get caught right away, you create only one account and open as many profiles as you can every second in your virtual browser. Each time you open a profile, that's called a profile view. Every website that worries about the costs of the data bandwidth they must pay for when people access their website, or that knows the value of their data, will monitor for extreme heavy users. Back when I worked at Google, my adventure catching Aaron Swartz was kicked off by automated baseline monitoring that watched for scraping of Google Books. Google didn't care about the bandwidth cost, but they cared about the money they had invested in scanning and processing the books, and they worried about people freeloading off that investment.

Facebook had limits built into the product to stop the crudest mechanical interactions with the website. Instead of asking, "What is a reasonable amount of usage for a human?" Facebook asked, "What amount of usage would hurt our systems?" As one Facebook document put it: Are you protecting people or machines? It's critical to remember that many stakeholders inside tech companies get bonuses and performance reviews that are built on metrics that lack true accountability. For example, there's a team inside Facebook that's rewarded for more "usage" of the profile. That team's interest is directly in conflict with the team responsible for stopping bots from downloading those same profiles. If, say, the upper limit for the number of profiles "someone" (or a bot) could view per minute at Facebook was thirty, a superscraper would come in right under around thirty profile views a minute to avoid suspicion.

Let's imagine one day the scraping team said, "No human looks at more than five profiles a minute." Or better yet, "No human looks at more than five profiles a minute for more than twelve hours a day, seven days a week." From that one super "user," 275,000 of their 300,000 weekly profile loads would disappear. If the scraping team does too good a job stopping bots, one could easily imagine 10 percent or more of the profile "usage" might disappear in the span of a month.

The Facebook efforts to detect and stop scrapers were real, but understaffed and underfunded, and that was intentional. Tech companies make a lion's share of their profits from advertising. If even just 1 percent of your users disappear when you report your quarterly earnings, your stock price could dive as much as 10 percent. Unlike the currency of brick and mortar businesses, there are no generally accepted accounting standards for "users" as there are for dollars, except those the companies define for themselves in their reports. A "person" can be as broadly defined as "an account who performs any action in our product" or "an email address we have." No company that reports how many users it has is required to reveal to auditors what it does to find and stop automated use. Instead, every company that reports how many users it has will be worth more if they have more "users" of all sorts — bots included — just as a brick and mortar store would be if they claimed they had more dollars than they actually have.

When I was at Harvard Business School, one of my favorite classes was Financial Reporting and Control. My professor did not teach the material as a series of rules to memorize, but rather presented the history of how and why the rules had emerged, and how and why they were necessary. We have accounting rules not only because companies and investors are more successful over the long term when everyone involved talks about money in a manner that is aligned with reality, but also because investors are willing to invest much more money when they can trust that a business is real and honest. That is, when truth and transparency are the cornerstones for evaluating the health of the business, both individual businesses and the larger economy benefit.

When businesses can cook their books to buy time or attract more capital, they'll do it. Instead of facing consequences early and being forced to either fix things or return to whatever the natural equilibrium with reality would be, they'll keep cooking the books while their problems increase until they're catastrophically harmful and the lie can no longer be maintained. I have talked to people who provide services to brand-name, smaller social platforms who saw firsthand that user counts fell by 90 percent when bots were adequately stopped. Given the opportunity to operate in the dark, those same companies turned down bot protections until the "users" returned. The platform clung to the idea that their users were real and the denial-of-service software they were using was broken and keeping their real users away. No one today gets to force them to face the truth. While 90 percent of Facebook's users are almost certainly not bots, I was about to learn about the institutional will to examine how much automated behavior was taking place, even when accompanied by a warning that a nation's information environment was at risk.

The case of the just under 300,000-profile-a-week superscraper shows the least intelligent way to design a scraper. I wouldn't be surprised if that account existed to see whether Facebook had anyone paying attention to scraping at all. If you penetration-test what the rate limit is and then hang out right below it, Facebook might very well allow the superscraper to keep scraping forever. It's like hiding your fraudulent cash transactions by keeping them just under the $10,000 limit that requires they be reported to the government.

If you wanted to hide, you'd choose to spread out your activity across a large number of accounts. But you'd still have to think carefully how to prevent your network of bots from exposing itself in other ways. The first big trade-off is how many accounts do you want to maintain versus how many profiles do you want to download? As you run more and more fake accounts to access the Facebook data you're interested in, you begin to have to hide any common infrastructure you're using. If you're using pure robotic accounts, you might have to spin up thousands or tens of thousands of simulated phones and computers, and scatter them around the world so that Facebook wouldn't notice too many "people" were using the same "phone" or that all of these "phones" were accessing the internet through the same access point in an outlier data center that generally didn't serve as a gateway for mobile traffic.

The question of "How many profiles do you want to download?" has a hidden requirement within it—how fresh do you need the data to be? In other words, how often do you want to download a new copy of each profile? If your business case requires up-to-date data, you might need to download a copy of a profile every day or maybe even multiple times a day. This adds another wrinkle. Every time your bot downloads a profile, it leaves another clue that your scraper network exists. One of the ways coordinated behavior is caught is by looking for multiple accounts interacting in consistent ways. In the case of misinformation, that might mean a set of accounts co-administering the same groups or all commenting on the same posts or posting the same links or focusing on information from the same set of domains.

To increase the chances of your scraping effort operating undetected, ideally your accounts would overlap as little as possible. If you needed ten profiles scraped, you'd have ten separate instances of your scraper look at those profiles. But now the number of accounts you need in order to cover all the profiles you care about explodes to a scale that isn't feasible to monitor. Throw all these factors together, and you have to start thinking about ways of optimizing your visits to profiles to try to hide your activity as much as possible.

One of those strategies is to have some of the repeat visits happen by the same account. Real people every day visit a single profile multiple times. Imagine someone still pining for their ex-partner. They might check in on their ex-partner's profile every day, or several times a day. Maybe some of your bots could visit the same profile each day so you could get your fresh data.

Now a new way to find your automated activity appears. Let's say I find a "user" who, for each profile they visit, looks at that profile the same number of times each day. Last week they looked at five hundred different profiles, but every profile was visited exactly seven times or maybe fourteen times (twice a day) or twenty-one times (three times a day). That "user" is almost certainly not a person. No one looks at every profile they encounter with the exact same cadence.

"Okay, Frances," you might say, "I know how to hide myself." Most of the profiles I look at, I'll look at once, fewer I'll look at twice, and even fewer I'll look at three times — now I'll be acting like a real person who looks at a bunch of profiles. But here's where it gets interesting. You don't really know how actual people behave. You have to make a guess for what that drop-off looks like. As a person in the business of data extraction, you have a vested interest in visiting as many profiles as possible. You'll almost certainly pick a coefficient (that drop-off from one visit to two visits, and from two visits to three visits...) that is radically higher than the coefficient that matches what real users do, because you have data to capture and business needs to meet. Lastly, and most importantly, because you can't see the real data, you're just going to have to make a guess for your coefficient value and run with it. Likely no one else will guess the same coefficient as you because it's just a guess. And if you use the same coefficient for all your bots, you just fingerprinted your network. Gotcha.

In the summer of 2020, as I was trying to figure out how to get rid of the noise in my model, I stumbled on this trick for finding at least some bot networks. I started clustering accounts that viewed multiple known Civic Actor profiles over the span of a week based on their revisitation patterns. Very clean networks began to pop out. And as I dug into those networks, something weird became evident. There were huge networks with *tens of thousands of accounts,* where each account was looking at thousands of profiles a week, and each account that individually and collectively screamed "I am not a person" was also clearly, simultaneously, a real person. How could this be? I remember sitting at my desk in my parents' sunroom, dappled Iowa summertime light streaming through the skylights, and thinking, "What in the world have I gotten myself in the middle of?"

The accounts that jumped out at me weren't new. I was focusing on accounts that had been made in 2007 or earlier. They had lots and lots of natural-looking behavior, like Messenger messages bouncing back and forth, "people" scrolling through the feeds and liking posts and clicking links, and yet they were looking at thousands of profiles a week in mechanically consistent ways (think those visitation thumbprints). Also, while all the accounts I was looking at were in the United States (our model was only focused on this country), a disproportionate number of the users had their interface language set to Vietnamese. Could this be a nation-state-run network?

I wrote up a report on what I had found and scheduled a meeting with the scraping team. They told me even the accounts that individually looked at tens of thousands of profiles a week did not meet the Facebook definition of scraping. I warned them we were potentially looking at a distributed threat — tens of thousands of accounts in just one network that could have software on their computers that was weaponizing their accounts. How could this *not* be scraping? They shrugged. As far as they were concerned, they knew the definition of scraping, and this wasn't scraping. They didn't say it out loud, but I knew Facebook's servers weren't being threatened by these accounts, and so, as far as the scraping team was concerned, it wasn't their problem. They only cared about the company's servers being able to handle all the activity. They told me if I was worried about this, I should talk to the Compromised Accounts team.

I reached out to that team with my report, and they asked for a sample of accounts to examine. We scheduled a follow-up meeting, and they told me after examining them, they didn't meet the definition of a compromised account. I don't know for sure, but I assume a compromised account in their world was an account that had been accessed against the will of the owner. If the people in the network were still genuinely conversing with their friends and there wasn't any outlier access (for example, divergent behavior happening in another country simultaneously), that account wasn't a compromised account under their approved mission scope.

I don't fault these teams for their lack of curiosity. Facebook was a place of metrics and clean, tightly scoped mission statements, and it seemed to me that both teams were radically understaffed. Uncovering new areas of harm doesn't advance your team if you don't think you're going to get more staffing to address those harms. On the contrary, taking on such work will just spread your team thinner and potentially compromise the metrics your performance reviews are dependent on. Think Sasha and his burnout. Everyone has a finite number of plates they can spin.

At this point, I wasn't sure what to do. I had been explaining to people what I thought the stakes were. If these were real people who seemed to be demonstrating that they had automated software on their computers or phones, it was easy to dismiss that software if it was just downloading profiles (even if the profiles belonged to known Civic Actors). But imagine if that software could also post new posts or comment on posts.

What if, a week before an election, even one of those tens-of-thousands-of-accounts-strong networks started pushing coherent lies? What if one of these networks were able to create the perception of a widespread false trend that was designed to shift the outcome of the election? How would the media respond? Would they even be able to get to the bottom of it before the election had come and gone? How would the public respond when the echo chamber came from everywhere all at once?

———————

Finally, I got some traction for our findings from the Threat Intelligence team. Part of an organization called I3 (Integrity, Investigations, and Intel), the threat researchers were experts who dug into Facebook's data to try to find bad actors. Many had backgrounds like service in the FBI, while others had been trained inside Facebook and had found their way into Threat Intelligence because it was the home for those who wanted to spend all day manually cleaning up the trickiest parts of the platform. So I met with the head of Threat Intelligence and a few people from Information Operations. If these networks flipped a switch and went from passively observing to actively influencing Facebook users, Information Operations would be responsible for the threat. They listened intently and said they would look into it.

My office was the mezzanine of my parents' home. My mom and dad could hear almost all of my meetings, at least when I happened to not wear headphones. By chance, they got to hear my shock when the scraping team told me what I saw wasn't scraping. They got to hear my questions as I sought to understand why exactly it wasn't scraping and hear the teams repeat almost formulaically that it was beyond their scope, and no, they had no idea which team would have appropriate jurisdiction.

It was these interactions that made me realize how special the Civic Integrity team was. Whenever I would talk with my colleagues there, if something appeared wrong, people would at least admit there was a problem, even if they had to deprioritize the issue because of limited resources. Other teams would nonchalantly give me a brush-off. It left me frustrated and scared. I now realize how fortunate I was to be locked down with my parents. I could decompress with them and temporarily escape from the troubling discoveries I was making.

Regardless of what went on during the week, on the weekends I could distract myself by cycling with my parents. At seventy-one and sixty-eight, both of my parents were still avid cyclists. Each weekend we'd throw our bikes in the van and take off to different rails-to-trails cycling paths that weaved through the hills of the rolling Iowa countryside. Starting in the 1960s and gaining steam in the 1980s when deregulation allowed a large number of unprofitable rail lines to close, community groups across the Midwest and later the rest of the country had been converting old railroad lines into trails so riders could enjoy the natural world. Iowa had a surprising amount of railroad capacity, a leftover of the era when the only way to get large-scale agricultural products to market was by train.

We'd head out in the morning, trying to beat the hottest part of the day. Those hours on my bike helped me unplug from the stress that was bound to flood back in once I sat down at my computer on Monday. I remember sitting beside our bikes eating lunch and wondering how many more times I'd get to share experiences like this with my dad, experiences that had been so essential to my childhood. It gave me the perspective to appreciate each moment we were out there together.

One of those adventures through the picturesque Iowa countryside took place in mid-August and went through the center of the state. We drove nearly two hours from Iowa City to Ankeny to ride the High Trestle Trail, one of the flagship routes in the state's rails-to-trails network. This ride, though, was different from our previous excursions. It passed through the center of the destruction wrought by the derecho that had torn through Iowa two weeks before.

A derecho is like a linear hurricane. Where hurricanes spin as they cross the ocean or move onto land, forming circular funnels with radial arms, derechos are a broad front of thunderstorms that sweep across the plains every few years. In 2020, however, the derecho that formed on August 10 swept across eight midwestern states with winds that peaked just twenty miles to the north of Iowa City at 140 miles per hour—the equivalent of a Category 4 hurricane. It's okay if you've never heard of a derecho before—many in Iowa had never heard that word before 2020. The winds recorded in Cedar Rapids were unprecedented, setting the record for the highest winds ever recorded in a derecho.

As we biked through the countryside, we saw the damage up close. Huge holes along the trail marked where trees had been ripped from the ground. In places, the trunks of severed trees flanked the trail on the left and right forming walls of wood debris, a physical manifestation of the love and care the locals had for their trail. Twenty-story grain elevators lay crumpled on the ground, their contents a graceful golden waterfall of corn not yet cleaned up and hauled off. It was the costliest recorded thunderstorm in United States history.

As the summer came to a close, Samidh started suggesting that I should try to find a new project. We had gotten our influence modeling to the point where it could find candidates within an 80 percent accuracy rate, but we were told we needed to be producing results that were 95 percent correct. By then I had started having virtual coffees with coworkers across Civic Integrity, partially to stay abreast of what was happening with the teams, but also because it expanded my pool of socialization after months of living with my parents. One day, a coffee with a relatively new addition to the team set me on a whole new path at Facebook.

Taylor had come from a senior role at Palantir Technologies, an established big-data analytics company with a national security focus, and she was now leading the product team that included Facebook's Coordinated Inauthentic Behavior and Information Operations teams. She had sat through one of my presentations regarding the scraping network and my concerns on how it might be repurposed. During one of our virtual chats, Taylor said, "You know, there might be an opportunity on my team." She informed me that the Counter-Espionage Threat Intelligence team was hiring their first product manager. She encouraged me to talk to them.

Within a few weeks my future after influence modeling was secured, and around the beginning of October I began onboarding onto the team.

Immediately it was like a breath of fresh air. The team was small, six or seven individual contributors and a manager. Some of them had been with Threat Intelligence for so long that they had been in the original combined Information Operations/Counter-Espionage team. In the beginning this combination made sense, as Information Operations often begin by surveilling a community or individual targets that might be influenceable. At some point, the Information Operations team had grown large enough that it made more sense to have two separate focus areas.

The Information Operations team had very constrained operational parameters. In the wake of the Cambridge Analytica disclosures, Mark Zuckerberg had gone before Congress and promised to report every influence operation found on Facebook, which gave each case much higher stakes. Espionage often was best dealt with by warning victims or making attackers less effective while leaving them on the platform — for example, by not delivering their messages. If we removed a bad actor, we would have to go hunt them down again when they rejoined under a different identity. We had cases that had been going on for years.

People on the Counter-Espionage team were much older, much more experienced, with more institutional knowledge, than on every other team I had encountered within Facebook. They reminded me of the days when I was happy at Google. People like to complain about how young the people at Google are, but Facebook was far more extreme. For context, at Google different subsets of employees had names that ended in "-ooglers". Employees in general were Googlers. If you were gay, you were a Gaygler. If you were new, you were a Noogler. If you were over forty, you were a Graygler. Facebook betrayed how young its employee base was with its employee resource group for older employees, a Workplace group called Facebook Seniors. A friend celebrated when she turned thirty and she could finally join — she said it made her feel like she was now an adult.

While age is just a number, it has real consequences with regard to organizational behavior. When your workforce has overwhelmingly come straight from college, leadership is more able to just assert, "This is the way the world is, accept it," and get compliance. I think the fact that I was over thirty-five, and that I had had enough time to get a Harvard MBA and work at multiple companies, played a critical role in my becoming a whistleblower. I knew that what Facebook presented as dilemmas were really false choices. My twenty-three-year-old peers did not.

It felt different to be on a team that was mostly thirtysomethings who were established in their careers. Here were people who were inherently mission-driven; several were veterans of the FBI, who every day showed up ready to make the world a little better place. They worked in virtual anonymity, knowing the work they did could potentially save lives, and that was more than enough reward for them. I felt at home.

Soon we were in the run-up to the 2020 presidential election. It may seem like the 2020 election is missing from this book, but that's because my day-to-day work life wasn't dramatically impacted by it, and I didn't have a lot of insight into what the US election preparations entailed until Stop the Steal and January 6 happened. Most of the actors we tracked for counterespionage were foreign actors targeting foreign actors. They were Chinese operations hunting Uyghurs around the world, compromising phones and using the photos on them to understand who they were and where and with whom they socialized. Or we were tracking Iranian catfishers flirting with Israeli soldiers. The world carried on as the US election swirled, and we were focused on that.

I have to admit that in October 2020 I relished the fact that we could turn our focus outward in this way. The last year had been intense, and I was happy to let someone else deal with the chaos of the last dash across the finish line. I felt I had already carried more than my share of that load, and I just wanted to keep my head down and finally get a chance to put down roots with a team at work with which I could potentially stay for a couple of years. I knew the vast majority of the Civic Integrity organization was focused on the election. I thought Facebook's election work was in good hands, even if more hands would be more ideal. Things felt like they were starting to look brighter. Oh, how little I knew.

CHAPTER 12

Go Time

If one is forever cautious, can one remain a human being?

— Aleksandr Solzhenitsyn, *The First Circle*

"**H**ow do I know you are who you say you are?" I typed into my phone. It was early December 2020, and I was responding to a message from someone identifying himself as Jeff Horwitz, a reporter with the *Wall Street Journal* who covered technology and who had written many well-reported and accurate stories about Facebook. This person identifying himself as Horwitz had originally sent a message to me through LinkedIn the day after the November 4 election. However, I didn't check my LinkedIn messages until a month later. I'm not much of a professional social networking user, particularly when I'm in a job, and yes, I know that goes against the advice of every career coach.

The LinkedIn note certainly seemed as if it were written by someone who had reported on Facebook, someone who had experience talking with Facebook employees struggling with internal conflict. "I'm sure you're exhausted," went the note. "I know things are still pretty volatile on all social media platforms.... I'd love to speak with you in the coming days about how the interventions made before and after the close of polls went, and what, if anything, of the election-related alterations to the platform might be worth permanently keeping."

Oh, the irony, I thought as I read the note. *If he only knew.*

———————

On December 2, less than thirty days after the November 2020 election, Samidh Chakrabarti posted a team message to everyone who worked under him in Civic Integrity. There were approximately three hundred of us, and we had done all we could to ensure that Facebook was not used to spread the sort of toxic misinformation or inflammatory content that might foment violence and bloodshed the day of the election or afterward.

That message informed us that Civic Integrity was being dissolved. In the grand Facebook tradition of managing through instability, the note Samidh had posted to the team's digital message board informed us that we were being "reorged." We would mostly be divided into three other teams that were working on adjacent issues. In his communication we also learned Samidh would no longer be affiliated with any of these efforts. After he led an all-hands conference call later in the day, he would be taking some time off. If there were any doubt as to how swift the end of Civic Integrity was, the very first message someone from the team sent in response to Samidh's announcement triggered an automated response: "I'm sorry but Samidh Chakrabarti is currently out of the office . . ." It seemed to me that Samidh was almost telegraphing his feelings regarding the end of Civic Integrity: He had posted the decision from on high, and he was immediately checking out.

I thought back on the impromptu Civic all-hands meeting he had called a little over a year before, in the wake of Mark Zuckerberg's decision that he could write a political-speech policy for the company over a single weekend, one that would be better than the one meticulously crafted by a cross-functional team of thirty people who had spent months examining every angle and considering the conflicting trade-offs. That was the meeting at which Samidh had shown us the MRI of his ruptured discs caused by too many hours at a desk, glued to a monitor, trying to fight the tide with just a thimble and a skeleton crew foolish or desperate enough to try.

Much later, I was told that Samidh had been informed a few days *before* the 2020 election that Facebook was going to disband Civic Integrity. I couldn't help but think that when he got this news he was in the middle of perhaps the most difficult period of his tenure at Facebook—in maximum exhaustion after months intensely focused on building election defenses with a team he had spent five years bringing together. It must have been incredibly demoralizing. Samidh and our team had made tremendous sacrifices in our quest to minimize Facebook's disruptions to American democracy. Whoever had pulled the trigger somehow thought Samidh should get the news just as we were racing to the finish line. Civic Integrity and all that he had put into it was going to be dissolved.

At the all-hands meeting Samidh held on December 2, he reassured us that while the reorg meant that there would no longer be a Civic Integrity organization per se, the company was going to continue investing in the "Civic Space," just in a different form. He tried to stay positive, comparing the Civic Integrity team to something like America itself—not only a place, but an idea. That, just as the idea of America attracted talented people from around the world, the ethos of the Civic Integrity team had made it more than a mere product group. He acknowledged that many of us had come to Facebook specifically to work on Civic Integrity, and perhaps more importantly, many had told him it was the first place they had experienced a true sense of kinship with others who believed in building platforms with a conscience, "doing Integrity work with integrity." Of course, Samidh understood why people felt uneasy or hurt about the reorg. The feeling of loss was real; there was no papering over it.

Samidh's words resonated with me. Over the years, I had declined to join Facebook many times because I wasn't sure where its corporate heart lay. The only reason I joined the company when I did was to be part of Civic Integrity. I thought back on some of the Facebook hires I was most impressed by and wondered if any of them would have come to Facebook if Civic Integrity hadn't existed.

In his address, Samidh talked about the intensity of the work our team had put into preparing for the 2020 election and that it was work we should all be proud of. Many of us viewed our efforts over the previous year as the best work of our careers. He noted that a reorg could feel like a "strange reward." Reorgs at Facebook usually had an obvious winner group and a loser group. All you had to do was look at who ran what when the dust settled. The fact we were being broken up into different subsystems and distributed meant there was no longer a center of mass. None of the three new leaders would report directly to Guy Rosen, the VP of Integrity, as Samidh had. It didn't seem like Civic was going to come out of the reorg with the same autonomy or authority that it had had going in. Samidh acknowledged that the work we had done had taken a toll on every one of us, taken a toll on "our family, our friends, and our health."

I sat there listening and tried to be open-minded when he suggested next steps — that our job was to go out into Facebook and cross-pollinate the values of Civic Integrity into the rest of the company. But now that Facebook had killed the very team responsible for Civic Integrity, the idea that we could go out into the company and carry that mission with us into our new assignments seemed divorced from the realities of organizational behavior.

Doing his best to inspire us, Samidh advised us that when we saw Facebook putting its short-term interests ahead of the long-term needs of the community, we should express ourselves constructively and respectfully. When we witnessed teams focusing too much on averages or prevalence, we should use our skills to look after and protect the most vulnerable. When we saw Facebook's platforms further entrenching the powerful, we should maintain faith in our company's stated mission of giving people power. If we were to do that, Samidh said, we could honor the legacy of what we had all built together on Civic Integrity. The way he put it, we still had a way to "live our civic duty."

This was probably the best pep talk Samidh could deliver given the circumstances, but in my fifteen months at Facebook I had learned that even within Civic Integrity, if you were seen as being too negative, you were viewed with skepticism and isolated. The few times I had ventured out to interact with other teams, the country-club vibe was even more intense. There was minimal space for criticism. Anything that rained on the idea that we at Facebook were doing the Lord's work of connecting the world was unwelcome. I had spent only a little over a month in I3, but part of why I loved it was that its threat researchers looked at Facebook's dark underbelly all day, every day, and there was zero need to spin overly generous interpretations of what was actually happening. Perhaps because so many of my I3 coworkers had come from professions in which the mission was to protect and serve, that team was convinced that you can't adequately protect if you refuse to see and address the darkness in the shadows.

At the all-hands, we were told about the three "pillars" we would be split into. (Bear with me as I use the company's org-chart jargon here.) The Foundations pillar made infrastructure for operating the integrity teams—things like the human operations platform that content moderators used, or the measurement and insights platform for understanding whether an experiment or classifier was working. The Ecosystems pillar built cross-problem tools like the software that ran the war room or detected content at risk of going viral. Then came the Problems pillar, which held discrete spaces where subject-matter expert teams lived. I, for example, was now within the Inauthentic Behavior Pod under the Problems pillar.

One by one, each new boss stepped up to talk about why they were excited about the reorg and what they hoped it would yield. In a demonstration that Facebook is an onion of layers, even in the transition talk I learned something new about the history of the Integrity org. One of the people who was stepping into a leadership role had been in charge of Pages Growth in 2016/2017, and he had first collaborated with Civic Integrity when they had formed a war room to deal with the Russian state misinformation organization. That Russian operation called itself the Internet Research Agency (IRA). Facebook had been caught flat-footed when the IRA built about a hundred Facebook Pages to reach half of the US population while spending only $10,000 on ads. I had heard about the relative pittance the Russian IRA had invested in that operation, and often the fact that they had bought only $10,000 of ads was cited as proof that Russia's online involvement in 2016 had been overhyped. I had never before heard that Facebook had been so concerned about Russia's effectiveness at building extensive audiences that the company had set up a war room to tease out what techniques had been deployed to reach so many people so cheaply.

In the Q&A that followed, I asked the last question. In setting it up, I said that a core reason why Civic Integrity had been founded was because Integrity at Facebook had in the past focused on "prevalence" and "scale" (working on problems that occur often) to the exclusion of focusing on high-severity but low-prevalence (serious but less frequent) problems. We all knew the consequences of ignoring those high-severity, low-prevalence problems. (I had gotten way better at speaking Facebookese.) I asked what was going to be done to make sure Civic wasn't just washed away by that cultural mismatch of priorities and got this for an answer: "That's a great question. We're hoping you'll all help us answer that going forward." I wasn't particularly encouraged.

The three leaders who were replacing Samidh all sounded upbeat and excited. One thing not obvious outside Facebook is how organizationally static it is. A large number of people, almost all the same age, had joined the company at about the same time, ten years earlier, and now there was no room for them to move up. The people above them were themselves stuck at the same corporate rung where they had been the past few years. Now that the mighty oak of Civic Integrity had been felled, there was a rare opportunity to branch out and take up the canopy space it had occupied. I wondered how much this contributed to other parts of the company sniping at Civic Integrity. It had been the only part of the company that had grown its head count in recent years. There was much to gain for leaders in more static parts of the company if Civic Integrity could be scavenged for parts.

What I did next is a ritual familiar to disaffected employees around the world: I opened up LinkedIn. As I scanned the messages, there was one with the subject "Hello from the Wall Street Journal." I knew of Jeff Horwitz. He had been doing some of the best reporting on Facebook's actions and inactions in India and their consequences. In his message to me, he included a contact number for Signal, an open-source encrypted messaging program that is so trusted that, according to the *Wall Street Journal,* many people in the US military and State Department use it for work matters despite this being counter to policy. That's when I opened up Signal to ask him, "How do I know you are who you say you are?"

––––––––––––

"Great first question!" he replied a few hours later. He directed me to his Twitter account and noted how it listed his Signal number. He told me that later that day he was going to post an article he had written in the *Wall Street Journal* and he told me what the title was going to be. An hour or so went by, and the article materialized in his stream of tweets.

My skepticism eased, and I continued our encrypted conversation. I told him Civic Integrity had been dissolved. He started asking me questions. How was it possible that Facebook could disband an organization they had already sunk four years and untold amounts of funding into? He had an appropriate level of skepticism that Facebook would write off that kind of intellectual and organizational capital. Why had they done it? I wasn't sure; they claimed that we were so valuable we needed to be incorporated into the rest of Integrity so they could learn from us. Did I believe them? No. Why else would both of our most senior leaders, the head of product and of engineering, leave Integrity?

His questions kept trickling in over the rest of the day. I signed off in the evening and we picked up again in the morning. He asked who else he could talk to to verify my story. I directed him to *The Information,* an online trade newspaper focusing on the tech sector, which had just dropped the first piece of reporting anywhere detailing the reorg: "Facebook Splits Up Unit at Center of Contested Election Decisions." The speed with which Alex Heath had gotten his article live contrasted with and in some ways illustrated the differences between his publication and the *WSJ,* and foretold what I would face in the future. Alex had likely learned about the breakup of Civic Integrity at the same time Jeff did and had probably jumped into action right away and started getting all the leads he could. For all I know, Jeff was busy doing final edits on his *WSJ* story when I messaged him, but the fact that he didn't ask for other people to verify my story until the next day showed a lack of pressure to go to press. Jeff could have had the scoop if he'd tapped his network of Facebook sources the day before, when we first were in touch.

I told Jeff there was way more that hadn't been reported, but I wasn't sure how to proceed. He responded that even slivers of information could be transformatively helpful to his reporting. Even if I were willing to just let him ask questions. Merely answering them could open up whole new corridors of research and focus. He said he would never pressure me to give him documents, and there was no rush. Jeff was based in Oakland, and we agreed to meet up that weekend in a park in the Oakland hills. It was about a forty-minute drive from where I lived.

I felt I was in a moment of inflection. What were my intentions with

Jeff? I had started thinking about what my obligations were with regard to Facebook since the summer, when I'd encountered so much willful ignorance about the decentralized scraping network. My parents and I had talked at length about what should be done. My dad had expressed complete disbelief that I lacked any obvious routes to escalate my concerns to someone who could help. He ran a clinical laboratory at a Veterans Affairs hospital in Iowa City, and he said even the lowest staff member on his team would see signs in break rooms, elevators, or even the lab itself, reading "Have you seen something that endangers patient safety? Call this number..."

The fact was, for months I'd been thinking about trying to get out the truth about what was and what was *not* going on within Facebook. There seemed a fairly obvious trade-off between when to contact the authorities and how much information could be collected. I suspected that as soon as I was in touch with anyone, anyone at all, to get out the facts, the risk of my identity being discovered increased exponentially.

Back in January 2020, when the disaster that was Facebook's insufficient preparation for the Iowa caucuses was an obvious train barreling down the tracks, I had a long conversation with a lawyer friend about what my options and rights were. He explained that I could provide information to Congress that I thought was needed for that body to meet its constitutional oversight duties. Also, he advised me that I could go to the SEC if there was evidence of the company lying, or to the Department of Justice if I felt the law had been violated. The FTC might also be an option, because it managed issues related to the internet, like privacy, and it had held Facebook accountable before.

He explained that there was a cottage industry of lawyers who help file SEC complaints on a contingency basis, and that with an SEC filing I could get federal whistleblower protections under the Sarbanes-Oxley and Dodd-Frank Acts. My friend, who was the general counsel for a start-up but was migrating toward being a full-time law professor, said that employment law and SEC law were not his areas of expertise, but that if I needed it he could direct me to people who could help.

For legal reasons I'm going to leave ambiguous the timing of a number of actions, including exactly when I began involving the lawyers I eventually came to primarily work with, and when my documents were shared with the SEC, Congress, and the *Wall Street Journal*. But you can safely assume that all of these things happened before the first article appeared in the *Journal* in September 2021. I can't share the legal opinions my lawyers provided without waiving attorney-client privilege, so I'm going to leave them out of the story until the summer before I went public, when the discussions turned to whether and how to disclose my identity as the Facebook whistleblower, as that advice wasn't legal in nature.

Back home in Iowa earlier that year, when I formed a quarantine "pod" with my parents, we spent many hours discussing the gap between what the public knew about Facebook's operations and what was really going on inside the company. It felt like, on the one hand, I was weighing incredibly high societal costs and on the other, potentially incredibly high personal stakes. Based on my time working with Civic Misinformation, and the clear trend lines between how Facebook operated in some of the most fragile places in the world and its growth rates, my sense was that there were at least ten to twenty million lives on the line globally. I feared that in places where Facebook had paid to become the face of the internet and yet had failed to provide basic safety systems for the languages spoken there, we would see future repeats of horrific violence like what had unfurled in Myanmar or Ethiopia. The problem was only getting worse.

I imagined tossing in bed in twenty years, unable to sleep because I knew I could have acted but didn't. On a pretty basic level, I felt that accepting that future was impossible — that to do nothing meant I was throwing away the second half of my life, awash in doubt and guilt. At the same time I faced real consequences for derailing the rest of my life by standing up for the truth now.

———————

The same summer I was agonizing over what, if anything, I would do to assist the public by exposing the truth about Facebook, I imagined a different way of life. Despite all my criticisms of opaque algorithmic recommender systems, the YouTube algorithm seemed to know exactly what my soul needed. In the evening I would zone out with a stream of YouTube videos on #VanLife conversion, tiny homes, and modern homesteading. I started asking myself how much I needed to live modestly but comfortably.

During the five years from when I'd gone back to work post-hospitalization to when I closed my much-reduced stock market bet (the downside of having held on to it for too long), I had built my savings from nearly zero to enough that I could conservatively take out maybe $20,000 to $30,000 a year for the foreseeable future. Definitely not enough to live in San Francisco. Even if I shared an apartment, there were no bedrooms to rent for less than $1,500 to $2,000 a month unless you had won the lottery by snagging a rent-controlled apartment. Even then, you put your future in the hands of whoever held the lease for your home.

YouTube painted a vision of a different way of life. I watched videos by Bob Wells, an American YouTuber, author, and advocate for minimalist, nomadic van-dwelling. Later I watched videos by people who had built more elaborate and comfortable vans. I thought, "I have an engineering degree. I can do this." I even started sketching out a van that I could live in someday with a partner and possibly even a child. The people in these videos were living joyful lives on the road while earning far less than $20,000 a year. Choosing not to be dependent on Big Tech didn't mean I had to figure out another way to fund a San Francisco life. I didn't need to live in San Francisco to be happy.

By that December, when Facebook shut down Civic Integrity and right about the time Jeff reached out, I had come to the conclusion that I was smart enough to figure out how to support myself on the internet. Maybe that would be freelance data-science consulting. Maybe it was entrepreneurship. I didn't need to know exactly what, I just felt confident I could get by. Reaching that belief in myself was a liberating inflection point for me. It allowed me to stop feeling scared about losing my job at Facebook, which actually made me better at my job, more resilient, more able to let things run off my back.

There's almost a Maslow-style hierarchy of needs your mind constructs when you're deciding whether to follow your heart. Now that the bottom of the pyramid—physiological needs—was covered, and the next rung—job security—wasn't an issue, I could aim for the higher realms of love and friendship, esteem, and self-actualization. But I had some practical issues to consider. If I did help alert the public, would Facebook come after me and decimate my savings with expensive lawsuits? Would my friends shun me? That one was pretty easy to dismiss: Most of my friends weren't at Facebook, and many of them had been skeptical when I'd joined the company. My friends would support me. This again was a major differentiating factor between my early twentysomething colleagues and myself. Many straight out of college built their whole social support system at work for the first few years—following their heart meant walking away from many of the people they knew.

The risk of Facebook suing me carried two consequences. One was the matter of my time, since being embroiled in a lawsuit can take up years of your life, years I would spend constantly second-guessing my words and actions in case they increased my risk. I had already come to the conclusion that I could support myself even if I was forced to start from zero again. After all, I'd already done it once, after my trek through the medical system. This meant that time was a primary threat of legal retaliation. Facebook had never sued a prior employee before, and for good reason. The optics of a (then) trillion-dollar company ganging up on a single individual were not good. Worse, if they sued me, I'd have the opportunity to take the stand and explain at length why I had done what I had done. This scenario seemed unlikely—Facebook had smart enough communications people that they wouldn't want to risk making a martyr out of me. Why roll the dice on giving me a larger platform? I didn't consider it a major risk.

That left the risk I actually cared about: alienating my dear friend, Simon, the person who had encouraged me to join Pinterest years before. Simon had been my most constant friend for over a decade, and I viewed him as a big brother. No matter whether I was at a peak or a dip of the roller coaster that had been my life over the years, Simon and his family had always stuck by me. Facebook, like many other big tech companies, has a system in which you need several advocates within the firm in order to get hired (a system that worsens diversity and inclusion issues, but that's a different topic). Simon had been one of my key boosters for the Civic Integrity job, and was by that point one of the more veteran employees at Facebook. He had followed a route that was nearly identical to mine from

Pinterest to Facebook. He had been a high performer whose wings were clipped because of "culture fit" issues.

If I took action against Facebook, what would Simon think of me? It seemed likely that negative press coverage would impact the Facebook stock price and Simon's family's economic security along with it. I thought about his two sons, Ethan and Ross, and how my relationship with them had inspired me and provided much-needed moments of joy during my deepest struggles with my regenerating nerves and the pain that accompanied that process. I thought of the magical moments, like teaching them to ski and taking them to the Fairyland amusement park at Lake Merritt, in Oakland. Was I jeopardizing my relationship with all of them if I stood up for what I believed in? If I lost Simon and his family by trying to get the truth out about what was going on behind Facebook's curtain of secrecy, I wasn't sure I could handle that. All of this was on my mind that weekend as I drove to a state park to meet with Jeff Horwitz.

Before I accepted the Civic Misinformation job I had consulted with multiple friends who were security researchers or consultants about the risks I might be exposed to if I joined the company. Every one of them was very straight with me. Foreign nations were using Facebook to wage information conflicts with their adversaries, and I was choosing to put myself in the middle of those conflicts. I should assume that my personal and work devices would be compromised at a minimum by Russia and China, and perhaps other countries.

Out of an abundance of caution, I asked Jeff if we could meet without our phones, somewhere secluded, and he suggested a state park in the Oakland Hills. I got in my trusty walnut-brown Prius station wagon and drove across the Bay Bridge, reminding myself that I was just going to hear him out. I didn't have to make any decisions that day. I navigated the switchbacking narrow roads into the hills and eventually came to the virtual pin Jeff had sent me that marked where to enter the park. I was a little early, so I sat behind the wheel, taking slow breaths and listening to some music.

At some point during my tenure at the company, I had watched an all-hands meeting where some executive, maybe even Mark Zuckerberg, lectured people on the "selfishness" of talking to journalists. According to the speaker, journalists themselves were selfish and just out to use Facebook employees' naiveté for their own glory. People who handed over documents or talked to the press were selfish because they did it for the short-term hit of feeling important or being able to brag to their friends about it. According to this message, all those selfish people accomplished was making the lives of their coworkers more difficult. As I listened, I was struck by the unstated assumption — that releasing the truth to the public would never change Facebook's behavior, that involvement with the outside world could never make things better, or different. That's the secret of those who don't want the world to change. When you feel there's no way things can change or get better, that's a sign that someone or something is distracting you. They don't want you looking for or noticing the fork in the road that means a different world is possible.

Now, sitting behind the wheel, trying to slow my heart rate, I thought about how that hypothetical narrative of interacting with the press differed from what I was actually experiencing. If I took this risk, I was hoping no one would ever find out it had been me. I guarantee you, most of the hundreds of other Facebook employees who have released documents to journalists never wanted anyone to find out it was them, and no one ever has. I just wanted to be able to sleep at night, and Facebook had stolen that from me by dissolving Civic Integrity. The best rationale I could concoct was that Facebook might be able to fix things on their own. But I knew that was never going to happen. This was not Plan B or C or D or E, or even F. It was down somewhere near Plan J or K. And now the next step of that plan was that I needed to open that car door and talk with this journalist.

In the end I was maybe ten minutes late in getting to my meeting with Jeff. It had taken me longer than I expected to negotiate with myself and find the courage to get on with it. When I arrived, he seemed unperturbed. I suspect he regularly gets stood up by potential sources, and that ten minutes late is practically on time for him. He carried a small bag and offered to carry the picnic blanket I had brought with me. We hiked into the park maybe fifteen or twenty minutes before finding an open, flat area off the forested trail. I spread out the blanket and we sat down to chat. Why, I asked him, hadn't more reporters written about the international implications of Facebook's behavior? What had inspired him to do so? How had Facebook responded? What did he think the stakes were?

He explained that Facebook had gotten a largely free pass on their international negligence because stories about people dying half a world away are hard to sell to many Americans. It was also hard to understand that the experience of Facebook could greatly differ from what Americans saw when they pulled their phones out of their pockets. Facebook over-spent to keep the United States safer than other markets. In America, claims that Facebook was leading to social violence sounded like science fiction. Jeff also told me that Facebook treated him pretty well. They had come to accept that he was the Energizer Bunny and they might as well treat him with respect because he was going to keep coming back over and over again. He kept going because he'd been following India's descent into social-media-facilitated violence for years. We were nowhere near close to hitting bottom, there or anywhere else.

I watched him as he spoke, trying to size him up. Was this the kind of selfish, self-aggrandizing journalist I'd been warned about? Before this point I had never interacted with journalists, except for the two and a half weeks on the Google associate product manager trip when Steven Levy had tagged along with us. Fifteen years later, the most vivid memory I had of Steven was that he had treated me with respect, as a human being, and he wasn't a twentysomething guy posturing with other young guys.

What kind of person was Jeff? Was he someone I would want to go on a long and difficult journey with? There was much to consider—was I in or was I out?

Jeff wrote for my first-choice publication, the *Wall Street Journal*. If I were to work with any publication I would have chosen the *Journal*, as it was viewed as center-right or even right in its political lean. I knew that if the *New York Times* were the first outlet to publish what Facebook was up to, the American right wing would automatically dismiss it out of hand as a conspiracy aimed at diminishing conservative power. The *Journal* was the most established paper in the country that met my criteria, and Jeff was the most senior reporter on its Facebook beat. But I needed to feel comfortable working with him.

Fortunately for both of us, I left the woods that day feeling confident I could collaborate with Jeff, even if I wasn't yet 100 percent sure I wanted to go all in. No worries, he said; I didn't have to rush my decision, and even if I decided to work with him, we could start slowly. Even simply having

Facebook explained from someone on the inside, or having someone to dig in deeper with when another source brought something to him, would be incredibly valuable.

———————

I went home to my empty apartment in San Francisco and my solo COVID-quarantined lifestyle. It seemed I had nothing but time to think about what I wanted to do next. I put off making a decision regarding Jeff for maybe a week. I'm sure I made excuses that I was busy with work or too tired and overwhelmed to deal with it. We were in transition at work as the remnants of Civic Integrity and our new host teams began to become accustomed to one another. While I had spent roughly six months at that point thinking about doing something like working with a journalist, it felt a little bit like I was contemplating Zeno's dichotomy paradox as I got closer to deciding it was actually Go Time. My situation was no longer hypothetical; it felt life-alteringly real.

Our lives are measured by the choices we make, and right about then, on a Friday, I was reminded in a most tragic way that we never know how much time we have left to make those choices. That day I got a video call from a friend, Jenna. I could hear her crying even before I could process her face on the screen. "Sean McCabe died!" she said. Jenna and Sean were close friends, both artists, and they had been among the core of my social life the previous few years. I was floored. It turned out that Sean's death was COVID-19-related — not due to an infection, but still closely tied to the pandemic and society's response to it.

Sean and I were very different personalities who lived wildly divergent lives, yet when our paths converged we were both profoundly changed. Back in November of 2018, I had met him for the first time when he let me and a mutual friend sit in on a conversation he was having with a different mutual friend about his struggles with PTSD. We had offered to excuse ourselves when we realized he was going deep, but he encouraged us to stay. What unfolded over the next fifteen or twenty minutes probably saved my life, because listening to his journey to overcome his struggles associated with the death of a sibling let me really accept how sick I was as I struggled to recover from the depths of my medical odyssey. I had received a PTSD diagnosis a month before but had fought against and minimized it as something I would treat "one day."

My path next crossed with his the following April at a party, when we sat on a couch by ourselves and started crying as we talked. I shared how his willingness to live in the truth had given me the strength to get my own life in order and get the treatment that I felt had cured me. I described the moment I had first felt joy again and how I had been shocked because it had been so long since I could access that emotion — and that I felt I owed that moment to him. We were both overwhelmed by the sense of relief we felt at being truly seen by another person. Sean had an incredibly big heart, and he knew what it was to struggle because he struggled every day. He invested so much energy in making people laugh because he knew the value of the fight to survive, to resist and live, despite the darkness ever pressing in on him.

And now he was gone. Part of what shocked me about Jenna's call was I thought Sean was doing better. Inspired by our conversation, he had gone in for treatment, and he told me by the fall of 2019 that he was feeling way better. Like many others, he had been worn down by the isolation COVID-19 imposed. Sean was an intensely extroverted person. He was most alive when he was on stage making people laugh or playing his guitar, and he was incredibly good at it. Lockdown had taken all of that away from him, and he had turned to his old vices to cope. And that, accidentally, had brought his heart and the amazing show that was his life to a stop.

Sean's death crystalized my resolve. Our lives are incredibly short. Each day is precious. Very little really matters. Having a large house in the Silicon Valley suburbs doesn't matter. Amassing the most points in your bank account among your circle of friends doesn't matter. Caring for each other and doing what we can to fight for each other — *that* matters. I stopped waffling. I knew what choice I had to make — I was in.

A few days later I bought a new laptop to use as my dedicated computer for what was coming next: making sure the truth about Facebook would one day see the light of day. I made sure never to connect it to the internet. I didn't know exactly what I was going to do, but I knew as I started to organize my thoughts and develop a plan that I needed to be extremely cautious. My security friends had warned me my personal or work phone or computer might possibly be compromised, but this laptop was off the grid, and I'd do what I could to keep it that way.

I pulled the laptop from its sleek Apple packaging and opened it for the first time. I worked through the setup flow, choosing my language and accessibility options, but froze when it asked me to enter my name. Jeff had told me after our last chat to think about what my pseudonym should be, and I hadn't known what to suggest. I've never had a nickname or an alter ego for use at parties or Burning Man. Even as a kid, I felt possessive about my name. I was a Frances, not a Fran or a Frannie. But now I needed to type a name into this computer. I typed in "Sean McCabe" and hit return. I don't like being the center of attention, preferring to lend support from behind the scenes. But Sean loved being the center of attention, and he was fearless. He wasn't using his name anymore, and I felt he'd approve of lending it to my cause.

That Saturday roughly seventy-five people came together at a park on the Albany Bulb, a peninsula, formerly a landfill, that jutted into the San Francisco Bay, for Sean's memorial. The organizers had warned us we might have to rapidly disperse since we lacked a permit and they weren't sure if a group of our size was allowed, even outdoors, during COVID-19 lockdown. We were on the outer tip of the peninsula and obscured from the road by a hill, but the cars parked along the road that led to the site were a clue that something was afoot.

I had been living alone for about two weeks at this point, after my return to San Francisco from Iowa, and the funeral was, ironically, a touch of the old days that highlighted the new normal I had found myself in. People told stories and played songs Sean had written. Someone cleverly had handed out wristbands of different colors that indicated how concerned the wearer was about COVID-19 — green if you were open to hugs, yellow if you preferred to stay at arm's length, red if you wanted people to keep their distance. Almost every person I encountered there was someone I hadn't seen in nearly a year, and I relished being able to hold each one close in a hug. People talked of how miraculous and short our lives are, and how Sean had burned bright every moment of his. That we should learn from his strengths by not being afraid to live as our true selves. To not deny our full embrace of life and our role in it.

I drove straight from the memorial service to Jeff's house in Oakland, and we had our first sit-down in his garden. "Are you in?" he asked. I told him I was. He asked if I had a pseudonym I preferred he use within the staff of the *Wall Street Journal,* and I told him he should refer to me as Sean McCabe. Seven or eight months later, when I met other *Journal* reporters and editors for the first time, they were stunned—they'd only ever known me as Sean, and they hadn't expected to meet a woman.

Jeff's wife had a talent for landscape design, and the giant succulents scattered among the rows of elaborate flower beds provided a tranquil contrast to the conversation that unfolded. I spent the next two hours answering Jeff's questions as he filled page after page of handwritten notes. We had begun. I wanted the world to know what Facebook knew—about spreading malignancy—and what it was failing to do to stop it.

December unfolded quietly from that point on. Our new teams within Facebook seemed to acknowledge that the powers that be had significantly diluted Civic Integrity by "integrating" our teams into new organizations. Facebook had a tendency to shut down for close to three weeks around the holidays, and many people were even more checked-out than usual after the reorg. I chose to work through the time people usually took off because there seemed no point in using vacation days to hang out at home alone. I remember spending part of the time examining search behavior to see if it could provide clues to potential espionage candidates. The project arose from the insight that if you wanted to do reconnaissance on Facebook, the best search engine was Search on Facebook. You might be able to avoid Facebook detection by trying to search for your target via Google, but only a subset of profiles seemed to be included in the Google results, and none of the more interesting social network–specific filters were available.

Almost immediately, patterns started emerging. The vast majority of people don't search on any given day, but people who are doing intelligence finding using Facebook do lots and lots of searches every day. Throw in characteristics like searching for keywords like "uyghur" or "uighur" and you could quickly find suspicious accounts that seemed to be methodically looking for the diaspora from China in the hotspots they had migrated to around the world. When I cross-referenced my results against known lists of suspicious accounts, my strategy of mining the search logs looked extremely promising. There were candidates from every major country we tracked who were exposing themselves through their search traffic.

When we sat down for our Espionage team planning in January, I was surprised that I was discouraged from pursuing it further. Espionage had perhaps six or seven full-time people on it, but there were already so many known cases that only a third of them could be pursued at any given time. It didn't matter that I had found a promising way to find new potential spies operating on our platform. We were already beyond capacity. We knew we were working only on the outer surface of a ball; what good would come from knowing it was a baseball or a beach ball?

I didn't push too hard on it because there were other projects I was supposed to be supporting, like detecting Adversarial Harmful Movements. In the wake of January 6 and the Stop the Steal movement, Facebook realized they had never built adequate defenses for what they called Adversarial Harmful Movements, or as they would later be called, Adversarial Harmful Networks. Facebook had for years seen movement-building taking place on the platform, wherein a small group of people would slowly grow over time and attract more support for their cause. What was new in the fall and winter of 2020, or perhaps not so much new as newly unearthed, was that there were people very clearly studying what the vulnerabilities of Facebook's Integrity teams were, and growing their movements in ways designed to exploit those vulnerabilities. Stop the Steal was a prime example.

Back in 2016, there had been a burgeoning movement of Hillary Clinton supporters that grew over the course of a couple of weeks into a single group—Pantsuit Nation. The politics of Pantsuit Nation and Stop the Steal were, of course, diametrically opposed. But what's more interesting is the variation in how the two groups grew. Pantsuit Nation was a closed group, meaning that you had to be invited to join. Stop the Steal wasn't a single group, but rather a network of different groups and pages. While a third of the people who joined Stop the Steal groups were self-joins (they searched for and joined a group), 67 percent came from invitations. Unlike with Pantsuit Nation, where, overwhelmingly, users joined because their real friends invited them, 30 percent of the invites to Stop the Steal groups came from only 0.3 percent of inviters. That's 3 out of 1,000 of people who invited someone. These people sent five hundred or more invites, and they weren't random unconnected people. On the contrary, they tended to be connected to other superinviters through interactions like commenting on, tagging, and sharing each others' content. In the top Stop the Steal group, there were 137 of these superinviters, 88 of whom were administrators of that or other Stop the Steal groups.

This was *not* an organically growing wave of people coming together on social media. This clearly was a movement, with leaders who understood how to grow on the platform. Seventy-three percent of these superinviters had bad "friending statistics," meaning that the inviters sent out volumes of regular friend invites, the first step to inviting a stranger to a group, and the people who received the friend requests perceived them as spam invitations because they rejected them at least 50 percent of the time. More concerning, but illustrative of their adversarial nature, 125 of the 137 hid the location from which they accessed Facebook.

Details like location obfuscation were part of the overall new trend in adversarial movements—movements that considered the people in Menlo Park the enemy. People would build up infrastructure that allowed them to rapidly rebuild their groups as soon as Facebook took them down. They would talk openly about what the game plan was for when they were eventually caught, or how they might delay that day. They knew they were violating the rules of the platform by spreading conspiratorial, violent, or hateful content. Instead of migrating to a platform that would welcome them, they were intentionally playing cat-and-mouse because operating on Facebook allowed them to pull new people into their movement who would never deliberately seek them out elsewhere.

Stop the Steal was not the only adversarial movement on Facebook or Instagram. Facebook was seeing similar movements appear in places as different as Germany and Ethiopia. In the latter there were horrific consequences, including ethnic violence. My role on the working group was more hands-off than on the other teams I'd worked on at Facebook. Here there were extremely good veteran data scientists already working in the space. My job was to work with Policy to make sure their questions were answered and that they had a translator between the data scientists' math and the implications of that math. Working together, we would draft criteria to define the difference between an adversarial movement and a movement that was welcome on the platform. One criterion might be something like whether the movement acted as if they knew they were in the wrong. Pantsuit Nation never discussed topics like "When we would be shut down," or set up a discord server they could all bookmark and run to if the group was shuttered. They didn't do those things because they didn't perceive they were doing anything wrong. Other criteria might include real-world consequences or which groups of people were being targeted with violence.

In February 2021 it felt like Facebook was racing to play catch-up after the violence at the Capitol on January 6. Researchers had autopsied the adversarial movement that had been used to help foment that attack on Congress, and it was clear in retrospect that there had been alarm bells ringing the whole time. Yet Facebook did very little about it until January 4, 2021, just two days before Congress convened to count the electoral votes, when the IPOC (war room) was created. As I said, Facebook was playing catch-up, reacting instead of taking potentially preventive action.

Between December 2, when Civic Integrity had been dissolved, and January 4, when enough people had returned from vacation to realize Facebook had let a small spark spread into a raging fire, there had been no one clearly tasked with watching the aftermath of the 2020 election and ensuring that the relatively peaceful election stayed that way. Safety measures that had been on for Election Day had been switched off. It appeared that none of the new leaders who headed up the remnants of Civic Integrity felt it was their responsibility, or maybe even felt able, to create a war room to manage the situation over the holidays.

In all fairness, this would have been a big ask. You'd be asking people who had been running at full speed for much of the previous year to stay on high alert during what was their first down period—you'd have to feel extremely confident you weren't crying wolf. What made it worse was that Facebook didn't reward the avoidance of potential disasters the way it rewarded putting out raging fires, to the point where internal documents fretted about "arsonist firefighters," who made careers out of putting out fires rather than preventing them in the first place. It wasn't possible to measure a crisis that had been averted, but in your performance review you could point to a horrific news cycle you had reined in with your heroic fix. On a more mundane note, it's also likely that the frontline operational groups that monitored the platform no longer knew the new chain of command well enough to even ask for an official crisis declaration. Civic Integrity was of no help because it no longer existed.

By 6 p.m. Eastern Standard Time on January 6, most of the "Break the Glass" safety measures that had been active for the 2020 election had still not been turned back on. Only twenty-four hours later would most of the measures be enabled. What would have happened had someone flipped the metaphorical switch for each intervention on December 15, or December 26, or even January 1? Unquestionably the adversarial groups would not have grown as fast as they had, and they likely would not have been allowed to spread the rage and lies or derail good-hearted patriotic people into a fantasy that would send some of them to jail for years.

Many of the safety measures were as simple as mandating that if your group got more than a certain number of hate-speech strikes, you had to appoint moderators to review all posts going forward. If that sounds like not much—as in, Won't the group's own moderators just approve their own posts?—it's almost certainly true. But the act of having to maintain even a modicum of self-moderation slows down the number of posts going live, because it's hard to find enough dedicated volunteers, especially in groups that don't have real communities that people are emotionally invested in. These kinds of interventions are known as "adding friction." People are willing to pay those micro costs for high-value content like a church's Facebook group or a cancer support group, but few are sufficiently inspired by shitposting that they will undertake the repetitive, boring labor of community-moderation. Asking a community to self-moderate improves the quality of content on the platform without centrally choosing which specific posts are "good" or "bad."

Other break-glass measures would have lowered the temperature by preventing Facebook's product from creating an artificial appearance of consensus. One such measure turns off commenting in groups that have repeated "strikes" for violence-inciting content on their posts. Let's say you clicked "Like" on a post or left a comment. Facebook was designed to return that post back up to the top of your news feed with each new comment that arrived. In the weeks and months after January 6, when many of the protesters who stormed the Capitol were questioned about why they did what they did that day, they talked about how it had seemed that the country really was under attack and that they'd felt a patriotic duty to step up and act. Part of that echo chamber was fueled by people piling onto Stop the Steal posts to emote their anger. Every comment forced it again to the top of people's feeds, retriggering patriots' urges to protect the country.

And then there were the safety measures that would have directly impacted what took place as the crowd stormed the Capitol. Facebook Live video, for example, was used by the crowd's leaders to coordinate the protesters as they swarmed, and Facebook had returned to the pre-election hyper-amplification of Live Video. This increased the priority of real-time videos (they were "boosted") so much that the streams almost always were sent to the top of people's feeds, even if they were only loosely relevant to them. Facebook did this even though they knew the company had a long history of not being able to prevent horrific content, such as suicides and murders, from appearing in Live Video. When the Christchurch, New Zealand, mass murder occurred, streamed live on Facebook, no one reported the video until 12 minutes after the stream ended. At least 4,000 people watched it before the police asked Facebook to take it down, in part because Facebook provided expedited content reviews only in the case of suicide, not murder. Facebook knew these boosts were dangerous enough that they reduced them for the 2020 election, but returned to the normal hyper-boosted state soon afterward. Treating Live Video more like regular video wouldn't have stopped January 6, but it would have made it harder for the leaders of the riot to coordinate or urge the crowd forward in real time as it surged into the heart of the Republic. Choices like this demonstrate how Facebook's ranking is not objective, but rather full of editorial choices, often made to further business interests like ensuring the success of live streaming as a product.

Facebook was moving quickly to develop tools for "disaggregating" networks — intervening in movements in subtle ways to cause them to stop growing or start shrinking. While many of my criticisms of Facebook are about Facebook not doing enough — largely about not providing equal levels of safety in most non-English languages, often with catastrophic results — this was one of the moments when I realized that Facebook's lack of transparency had consequences on both ends of the effort scale.

Researchers at Facebook digging into Stop the Steal and the emergent Patriot Party had learned a great deal about how adversarial networks grow. Each of those networks had a small number of core members that actively reached out along what was dubbed "information corridors," such as adjacent affinity groups or topics, to find people who might be persuaded to join the movement. What if you just stepped in and excised those accelerators from the network? Or, maybe more subtly, demoted their posts or limited how many people they could interact with on any given day? In the case of interventions like forcing posts to be self-moderated or turning off comments, or maybe even turning down the amplification on a featured Live Video, a member or a leader of a network would notice something had changed and might try to counteract it. At a minimum they would be able to complain to the press or the government about unequal treatment. But the new, more subtle interventions would likely be invisible. All you would see is that your group had stopped growing.

As I began to sit in on meetings about disaggregating networks, I started to feel deeply uncomfortable with the new scale of power my coworkers were talking about. Facebook was building tools that in invisible ways would allow it to choose which movements lived or died. This was a power that should not exist without public accountability for how it was used. History has taught us that unchecked power is inevitably abused. For example, I would be unsurprised if Facebook reduced or blocked the distribution of my posts or suppressed any groups I organized on the platform. It's too hard to resist touching all knobs once you give yourself the power.

———————

I had been considering all these revelations alone, keeping them to myself. I'd spent almost all of December and most of January living alone, and it was starting to wear on me. Every day was the same in COVID-19-quarantined San Francisco. I lived by myself, I worked online all day, and then I watched Netflix or YouTube in the evening. Maybe if I was lucky someone would be willing to hang out in a park six feet away from me, though many of my friends' "bubbles" had rules that looked askance even at that. My neuropathy was far better in February 2021 than it had been in 2017 or 2018, but it was in no way healed. I had graduated from a cocktail of two neuropathy meds to three the summer before, when my mom had sent me a journal article about low-dose naltrexone, and it meant I went from being dead-tired at six o'clock every day, too exhausted to do anything other than melt into a couch or go to bed early, to having three hours after work that I could decide how to use.

While it was a life-changing improvement, having a couple of hours of free time a day isn't the same as being better. Every day meant a choice between being lonely or being out in the damp chill of the famous San Francisco fog, the cold-induced aching in my legs impossible to avoid. It was decidedly grinding me down. In an effort to try something different, at the start of February I moved in with two friends in the Oakland Hills, which helped somewhat, but I still felt lonely. It was a far nicer place than where I lived in San Francisco, and the view of the bay took my breath away when I would watch the sunsets as the fog rolled in from the Pacific Ocean. But I still felt very isolated. I was now a fifteen- or twenty-minute drive from hanging out with others, and I felt like I was rattling around in the large house. It was due to be renovated in a year, when modern insulated windows would be put in, which meant we had to keep the thermostat low to rein in the heating bills. I never felt warm or comfortable unless I was under an electric blanket.

It was in this context that I felt a certain excited curiosity at the messages gliding across my phone screen. I had been invited into a group chat because I'd been an early investor in a project. Within a few days of joining the other investors I started seeing little messages shooting back and forth about people moving to Puerto Rico. I dropped into the direct messages of one of my friends in the chat and typed "Excuse me, what's going on?" The scattering of people I'd known who had moved to PR before COVID-19 all seemed to be largely the young retired, not people who were still trying to accomplish things.

"Oh," he responded, "there's like six people in this chat that are moving to Puerto Rico this month. We're tired of San Francisco being closed and always being cold."

The wheels in my head began to turn. I hadn't been on vacation in almost two years. I had a huge amount of time saved up, and I hadn't been to Puerto Rico since 2007. Why not?

I got on a plane a few weeks later for a three-week trip that I knew would become permanent the very first day I walked into the ocean and heard the waves. I love the sound of waves so much I have multiple Spotify playlists dedicated to wave sounds. The only nonprescription treatment that lessens the pain in my legs is warmth. How had I never considered living in a warm place near the ocean? How had I gotten so blinkered on living in San Francisco that no other life seemed plausible?

I sank down into the water and tilted my head back, my body bobbing as I floated atop each gentle wave as it neared the shore. I felt relaxed. I felt like my feet and my legs were part of my body again. I felt whole.

CHAPTER 13

Extraction

Archive as if the future depends on it.

—Lisbet Tellefsen

"Frances, I'm disappointed in you." The words sliced into me. It was March 2021, and I had let down my new manager, Rahul. And for the first time in a year, I actually cared when my manager voiced disapproval. In Facebook parlance, Rahul "supported" the product and program managers in the newly formed Coordinated Inauthentic Behavior pod, which is where I found myself after Civic Integrity had been dissolved. But this job description fails to convey all of what Rahul meant, both to the team and to me. He was without a doubt the best people manager I had crossed paths with in the nearly ten years since I had worked on Google+.

Facebook's prowess with misleading nomenclature even filtered down into the names of individual teams. There was no ethnic violence or genocide team—it was the "social cohesion" team. As in, when social cohesion breaks down, ethnic violence happens. You weren't choosing which social movements to kill or stymie, you were "disaggregating" them. No one was compulsively using Facebook or doomscrolling, they were merely exhibiting "problematic use."

My new pod had a straightforward name in comparison. We were six to eight product managers and program managers who focused on people acting inauthentically on the platform. We formed the Inauthentic Behavior Pod within the Problems Pillar. Some of my peers focused on things like spam while others worked on Coordinated Inauthentic Behavior (say, a group of accounts playing both sides by posting to both conservative and liberal groups). Two of us were anomalies in that we were positioned within Central Integrity, while the teams we worked with were all threat researchers inside I3. We straddled these two worlds because there weren't enough I3 product managers to merit a manager for such a small domain. My focus was on counterespionage while the other product manager, Mike, worked on information operations.

When Guy Rosen announced the reorganization of Civic Integrity in a Workplace post, he explained what differentiated Civic Integrity from the traditional Integrity teams, which "focused (by design) on execution excellence within defined areas." To generalize, they had clean metrics that they steadily improved. His wording almost implied that Civic Integrity lacked "execution excellence." He acknowledged, though, that Civic Integrity "pioneered critical innovation" by taking "a broad end-to-end view of Integrity, understanding how the sum of our systems operate[s] and what falls through the cracks."

In January 2021 Mike and I would discover what pursuing "execution excellence" would look like. In the run-up to the 2021 First Half planning cycle, Rahul had informed the other I3 product manager and me that the domains we were covering—Information Operations and Counter-Espionage—were not "big enough." Mike and I had kicked off the planning cycle by eliciting feedback from all the leaders of all the individual teams on I3. We asked them what they considered their most pressing needs in order for them to accomplish their goals and learned that one of the two areas that most needed attention was "detection." Most teams within I3 had devised code for trying to detect the bad actors on Facebook that fell within their domain. Their code might detect an adult who was a little too interested in children, particularly vulnerable ones. Or a terrorist cell recruiting new members. Or an Iranian-run information campaign. But it was all extremely manual and as a result labor-intensive. When it came to rediscovering bad actors who had been kicked off the platform and had returned—recidivists—huge amounts of labor were spent on tasks that could likely be automated, albeit with some effort.

Soon after I started working on detection infrastructure, Mike rotated off of I3 into the virtual reality portion of Facebook. Mike had been clear with me when he explained why he was transferring. If Information Operations wasn't a big enough domain, he said, Central Integrity didn't really understand or value the threat that foreign Information Operations posed to the company or society at large, and he didn't see any point in trying to convince our new boss that he added "sufficient" value. I didn't have the luxury of rotating to another team. I was still on the long tail of probation from my time with Civic Misinformation, and I had to make my new situation under Rahul work. I would still contribute to the Counter-Espionage team, but I would also support the teams across the entire I3 group when it came to defining a coherent strategy around detection systems.

I interviewed each subgroup to learn their processes and needs for proactive detection (finding new candidate threats) and recidivism detection (finding past targets who had at least once already been removed but were trying to sneak back onto the platform).

Rahul's disappointment sprang from a gap between what I had promised and what I'd delivered. I was overdue in giving him a summary of what I had found and a framework for thinking about how to approach detection tooling. I told him I had one holdout team that kept canceling on me and was only able to finish the last interview the day before I turned in my summary. That's when he delivered that line about his disappointment. I felt like I was in elementary school again and I had forgotten my homework. I genuinely believed that Rahul cared about my success. I wanted to make him proud, and I had let him down. He told me he wasn't disappointed in me because I was late; rather he was disappointed in me because I hadn't been brave enough to tell him I needed help.

He looked at me with the compassion of someone who has considerable experience coaching others in becoming their better selves, and said something that would change my life: "We solve problems together. We don't solve problems alone."

That simple observation has become an anchor in my interpersonal relationships ever since, and even colored how I viewed Facebook flailing in secret versus seeking help from the public. There's a tendency among high performers and those who may not have had their needs met early enough in life to suffer in silence rather than ask for help. Was your past success a fluke? If you don't have a track record of asking for and receiving help, it may be difficult for you to believe you're worthy of it. Or that you can trust others to provide it.

One of the many things that stood out about Rahul as a manager was his willingness to accept me where I was while simultaneously coaching me and challenging me to get better. It made me wonder what my day-to-day at Facebook would have been like had I pulled the trigger back in my third or fourth week on Civic Misinformation, when I had realized we were playing with the deck stacked against us, and transferred to some-where else inside Facebook with a manager like Rahul. Would I have found myself documenting for history some of Facebook's misdeeds? Would I have found a way to make just enough impact that I could have continued to rationalize grinding along for another week, another month? Driving enough impact that I could excuse the flood of misinformation or violence incitement building up on the other side of the dike because I was patching just enough leaks?

But I didn't rotate out to some place better grafted into the trunk of Facebook. I had continued to push for change from the end of a rickety branch alongside my other Civic Integrity peers, and now even that branch was gone.

Moving to Puerto Rico had given me a second (or third) wind at Face-book. The tropical warmth wrapped itself around my legs like protective socks, easing my pain enough that I could reduce my daytime neuropathy meds by half. I was now living with two friends in a condo in San Juan, located on the northern coast, facing the Atlantic. The condo wasn't on the beach but was high enough and close enough that it afforded a striking view of the sea and the sky. The future seemed as limitless as the horizon. I could breathe. Anytime I'd feel myself flagging during the day, I could walk out onto our patio and soak up the view. When the tropical sunshine had melted enough of my stress, I'd return to my laptop to push onward.

With Rahul's help, and now embedded in the more traditional part of Integrity at Facebook, I was finally picking up on aspects of Facebook's way of doing things that I'd never grasped before. For the first time in my ten-ure, I really felt the wind at my back. I had a sense of security. And yet, as I continued to answer Jeff's questions for the *Wall Street Journal,* I knew I was destroying that security. I was documenting for history what I viewed as

Facebook's legacy. Over the winter and early spring, between my December meeting with Jeff in the Oakland foothills and taking up residence in Puerto Rico, I had time to think through how I could copy the documentary proof of what I had encountered in my job—covering both what was and what was not really going on inside the company.

For legal reasons, I won't detail the thought process I went through when deciding how, when, and to whom I'd disclose the documents exposing Facebook's wrongdoing. But I can walk you through some of the factors anyone undertaking a project like this might consider.

———————

Generally speaking, my efforts to extract information from Facebook were analogous to what the decentralized scraping networks I had found the summer before had been doing. At least, that was my frame of reference. An employee who is contemplating documenting an employer's behavior for the public good is constrained in many of the same ways a state agent or information broker is constrained when downloading profiles from Facebook. In both cases, one party is extracting information and the other is (ideally) preventing that extraction. The main difference is that the state agent collects individual people's data and the employee collects the corporation's data. The extractor must be able to articulate their goals and priorities if they wish to achieve them, because a thoughtful defender won't make success easy.

The decentralized scrapers I had stumbled upon the year before wanted fresh data, and so they hit the same profiles multiple times; they wanted to discover changes to profiles soon after those changes occurred. They were focused more on "depth" (multiple visits) versus "breadth" because they weren't trying to hit every profile on Facebook. Instead, they appeared to care more about civic actors. An employee inside the company might need to make similar strategic decisions. Is their goal to shine light on a single issue in great detail? Or illuminate a broad swath of behavior? Do they care about the most current problems, or about the series of actions that led to the current state?

The company also has goals, at least implicitly if not explicitly. In the case of Facebook's approach to scrapers, they had to consider how often they were hampering legitimate usage or how many staffers to devote to trying to catch them. Or, perhaps less nobly, how much usage of their website do they want to lose when they report their quarterly performance? Traffic is a valuable metric, even if some of that traffic is deeply problematic. How do they weigh all of that against the potential risk of a publicity incident like Cambridge Analytica, when Facebook paid a $5 billion fine and locked itself into the hot seat with the FTC for years to come? Most of the trade-offs are about short-term concrete costs versus long-term hypothetical risks.

In Facebook's defense, it can be incredibly hard to weigh one of those against the other. In the process of interviewing all the teams in the I3 organization, I had been shocked to find that none of them had robust systems for proactive detection. More shockingly, when it came to recidivism, teams were working with the most rudimentary of tools. As I dug deeper into detection, I came to believe that when Facebook considered the long-term hypothetical costs of things like detecting and curtailing potential scandals (for example, over human or organ trafficking, terrorist organization activity, drug cartel activity, large-scale fraud, scraping operations, child exploitation and endangerment, or espionage), the company decided it simply wasn't worth paying the short-term costs of additional staff and safety measures.

Part of why I'm so tireless in advocating for public transparency and accountability — particularly for a company like Facebook, which has such rapid staff turnover — is that future risks will be paid by future employees, but short-term costs have to be justified and advocated for by current employees. Facebook will only adequately invest in these vital national security systems if they're audited and held accountable by the public. Keeping corporate documents secure is also not free. When it comes to a corporation trying to protect its documents and information, consider that high-sensitivity organizations, like the Department of Defense or various contractors within the military-industrial complex, know how to minimize the chance of extraction to near zero, but the process ends up being incredibly invasive while also being a drain on productivity and budgets. They do things like prohibit people from taking their work devices home at night. This might not seem like a huge cost, but you end up losing out on evening work the company might otherwise have been able to gain from employees, and you make all the parents and others who need work-life flexibility unhappy. And as Edward Snowden demonstrated, even these steps can fail to stop a determined employee who wants to expose the truth.

A high-sensitivity organization will have to search or screen employees to ensure that their devices don't leave with them, which makes employees feel they're not trusted and results in friction every time they leave work. You have to pay for significant staffing to do those searches, because top talent won't tolerate widely variable wait times to get through screening; they'll accept missing their buses due to security delays only a handful of times before they'll look for another job (and tell their friends to avoid you as an employer). You also have to hire a staff that's tasked with looking for anomalous usage like higher consumption of documents or accessing older documents. They'll find many, many more false positives than true positives, statistically most likely from your most diligent and hardest-working employees. Which kind of employee do you prefer? One who doesn't build on the work of their peers out of fear of having to defend using older documents or one who seeks to understand how your business operates so they can help push it forward?

One option is to do something like what the Transportation Security Administration does — institute a big, showy, very public series of measures, even if in fact those measures fail to detect 95 percent of dangerous materials. At least the messaging of the security theater is clear: "We will catch you." Facebook's version of this was to make transparent to its employees some of the defenses it used, in the loudest way possible. Whenever there was a major unflattering news story about the company, Facebook would check to see who had last accessed the relevant documents mentioned in the article. If, during that month, only one person accessed a document that was two years old, and that document was included in the story, the company would make a public show of firing that person, and would make sure all remaining employees knew about it. They would bring up the firing over and over again and tell people, "This will be you if you cross the line." But those announcements were relatively few and far between compared to the constant flow of stories about Facebook. Most of the documents mentioned in such media reports had not been widely distributed, but still each had been seen by tens or hundreds of people in the previous month or two—almost always too many to effectively root out who shared them. Only if a series of stories came out, or if a single story used multiple articles for evidence, would they be able to determine that the common thread across all those files was a single person.

Other ways Facebook monitored us were more subtle. One day a couple months after I joined Civic Misinformation I needed to document how a reporting flow worked inside the live Facebook platform because we were going to suggest changing it. After I took six or seven screenshots, each representing a step in filling out the form, I was prevented from being able to copy/paste what had been on my screen into the document I was compiling. It appeared the company had spyware on our laptops watching for people trying to accumulate screenshots and blocking that behavior.

Some defenses make perfect sense. Over the five to ten years before I joined Facebook, there had been a number of scandals involving employees who accessed the private information of individual Facebook users to stalk prospective, current, or former romantic partners. These scandals scaled from people looking up whether a woman they met at a party was single, to women switching hotels on vacation to get distance from their Facebook-employee partners, only to have the employee look up the internet computer address the woman was accessing Facebook from and show up on her doorstep. After years of insisting that they took privacy seriously, these scandals were too damaging to be allowed to continue. Facebook had responded by building rings of trip-wires around private data that gently warned people away from fireable offenses. I regularly ran into them when I was working on civic influence modeling.

I had several acquaintances from my time at Harvard and Google who had either run for office or become affiliated with political campaigns. This meant that I was now often "two hops," or friends-of-friends distance, away from the very people I was identifying via civic influence. Once every few days I would spot-check the profile of a candidate my algorithm had identified as a civic actor and I would be told I couldn't learn anything about them because we were too closely connected. I could view their information, but my access would be reported to my manager and corporate security, and I should be prepared to defend why I had accessed it. I always declined to check that candidate my system had generated; they had made the data difficult enough to view that I was discouraged from using it for legitimate business purposes.

The safest extraction strategies an employee could use should look like their "organic" everyday behavior. They could use a photo camera to take moderate-resolution photos of their computer screen (4-megapixel photos will capture most of the information in a given document, and using higher resolution won't gain anything except on highly detailed graphics or charts). That would provide an "air gap" between the info and the employee's behavior on the company's laptop. If, instead, the employee printed to PDF or saved the HTML of those files, that might be a monitored behavior like the screenshots I was taking to document the reporting flow. Back when I was doing that project, I stopped documenting the flow. I now knew I had software on my laptop that blocked that behavior, but did it also report it to corporate security? If I'd persisted in doing it again, would someone have asked me to explain what I was up to? On the other hand, if an employee wanted to take photos of their screen with a phone, as long as their laptop camera was covered up, there was no easy way to detect it.

Then there are access patterns to be considered. Let's imagine that the scraper from the year before had wanted to be perfectly invisible, and had installed their software on the computers of tens of thousands of people. If they captured only the profiles that those people visited as part of their day-to-day usage of Facebook, it would be impossible to differentiate their scraping from the actions of average users because it would literally be average behavior. I was only able to find the rogue networks because they were looking at specific profiles over and over again in unnatural and consistent ways.

An employee who wants to be invisible doesn't seek out specific documents beyond the scope of their job, or at least not often. They could take photos of documents they had already opened as part of their job, maybe by simply accumulating more open tabs until they went home. Each night as they finished up their work at home, they could go through and photograph their tabs before closing them. Even this behavior might be detectable — some websites ping back data on how quickly documents are read. Going through each one too quickly (or doing this every night religiously) might throw up a red flag.

Protecting against the extraction of information is always in tension with usability and productivity. Maybe that employee who seemed to reread a number of documents at the end of the day really was trying to build enough time into their day to think deeply about their work. Research has shown that creativity needs "white space" to really breathe. Employers that treat employees who work the most hours (or act the most diligently) with suspicion also threaten their ability to retain the engine of their own success.

One issue I agonized over was a big one: when to let anyone else have access to what I collected. I couldn't get whistleblower protections until I submitted a filing with the SEC, but as soon as I did that, it would expand the circle of trust and add risk to my ability to complete my accountability project. With Jeff and the *Wall Street Journal,* if I were to give him the documents, I could ask him not to publish stories based on them so the company wouldn't know until I was finished. But what about Congress? To maximize my legal protections by passing documents through Congress first, I would have to trust congressional staffers to keep everything secret.

Meanwhile, I was making a map in my mind and on my clean laptop of what I thought, based on my firsthand experience, were the most significant lies and omissions in Facebook's telling of its story to the public. I was scared about the process and consequences of documenting Facebook's misdeeds, but I also wanted to keep focused on why they mattered and how they needed to be presented to government regulators who could act on the information alongside the public.

I had already seen one consequence of the many other Facebook whistleblowers. Each time a really damaging story was published, internal collaboration groups that had been open to the broader Facebook community became closed. I knew the public needed a full picture, and I worried that if I were caught before I could finish, Facebook would close any doors I had walked through so that no one else would be able to attempt it again. If the public was going to get an adequate picture of what was happening inside

Facebook, I had to get everything critical on the first try, in one scoop, one big highly manual scrape.

I had to get enough information that the public could help Facebook save itself. Companies struggle alone with their lies instead of solving problems together with the public for the same reason I hid from my manager or why I struggled alone for so much of my twenties — because they don't believe they deserve or will get help. And just as individuals can fall into closed loops of thought when they struggle, Facebook locked itself into spirals that were hard to exit once they'd begun. Facebook lied to the public because they didn't want to spend the money necessary to keep the platform safe. Stock analysts were endlessly hounding them for every penny of profit, but none asked about the costs external to Facebook that accompanied the ever greater returns. Facebook didn't want to involve other stakeholders in resolving trade-offs. They wanted to move fast and use that extra money to drive up their stock price.

Unfortunately, you can't hide from the truth forever, no matter how convinced you are of your own invincibility. When the truth broke, Facebook lost users and advertisers, and had to dramatically increase their safety spending. Facebook's stock price peaked at $378 the Friday before the Monday when the first *Wall Street Journal* article appeared. At the time I'm writing this, it's hovering around $120. Less than two months ago, in November 2022, it was below $90 — more than a 75 percent decline. My hope is that we as a society will decide transparency is just good business. I'm sure investors would have preferred slightly slower growth than watching their retirement accounts go up in smoke.

––––––––

Part of what helped me navigate the stress during this time was a new person in my life who was providing a base of support unlike anything I had encountered in years. When I started my visit to Puerto Rico, of the five or six people I knew who were moving there, I only knew one of them as a friend. We were two of the first to arrive, and I ended up having dinner with him something like sixteen times in the first three weeks because I didn't know anyone else. We just clicked in a way I hadn't encountered many times before in my life. Chris was smart, kind, funny, and we shared a broad overlap of interests. He was tight with some of my closest friends back in San Francisco, and they vouched for him. Every one of them assured me he was a person you could trust.

We moved in together after maybe a month of knowing each other because housing in San Juan during the COVID-19 lockdown was impossible to come by. I was far from the only person fleeing a cold and gloomy place to spend more time outdoors in the tropical sunshine. I had been paying $3,000 a month for a tiny studio only slightly larger than my full-size mattress. At least the new condo had three bedrooms; I had only accepted sharing the condo with him and another friend from the group chat because I could have my own room.

I had opened up to him about what I thought Facebook was doing and why the public needed to know, and told him about my work with Jeff. He stood behind me and with me, and did what he could to care for me when the stress of it all became too much. He would plan weekend trips out to beaches beyond San Juan where all you could hear was the roar of the surf. I would swim in the ocean and feel held by the water and sun. It felt like that moment when I was crossing Utah as I escaped San Francisco. The permanence and soothing beauty of nature made the work drama of my job and my efforts to expose Facebook's wrongdoing seem ephemeral. The warm water and jewel-toned sunsets of the Puerto Rican beaches reassured me that nothing really mattered beyond the moment I was living in — until I would dive back into the fray on Monday morning.

One weekend the trip he planned for us and our roommate was out to Fajardo, on the northeast corner of the island. Another one of the people from the San Francisco group chat had just moved to the island, and his parents had rented a house so they could visit him. Alex and my then boyfriend, Chris, had been friends for over a decade in San Francisco, and now they were continuing that friendship under sunnier skies.

Fajardo is home to the largest marina in the Caribbean, and as we sat on the roof deck of Alex's parents' vacation rental and enjoyed a barbecue, we had a magnificent view of the blue ocean extending to the horizon. I could see the charm of buying an apartment in Fajardo and spending my afternoons on a small sailboat. My daydreams of a tranquil future by the sea were interrupted by the persistent buzz of a small quadcopter drone.

Given the security concerns I had been considering, the buzzing drone unnerved me. I could feel my anxiety building as it hovered just beyond the edge of the roof deck, maybe thirty feet away. I knew the drone almost certainly had nothing to do with Facebook, but I still was irrationally afraid that they had come for me.

The drone stayed there, its lack of movement making its intentions seem even more ominous. I could sense Chris and our roommate growing equally if not more tense. I wondered what thoughts were running through their minds. Alex's dad, Bruce, sat across from us, a fat half-smoked joint between his fingers, looking unfazed as our shoulders kept creeping closer to our ears and the tension built. He was a decisive man who exuded confidence. He had done everything from running a department store to driving a cab to developing international real estate. He had experienced a roller coaster of a life with extreme highs and lows, and he possessed the air of someone who believed he could overcome any obstacle. And he was an incredibly protective father. Bruce looked at the three of us. We were not his children but we were his son's friends, and that meant he had to do something. He could tell the drone was getting to us.

Suddenly he slammed his hands down on the table. "Fuck this shit," he said as he got up and walked maybe thirty feet across the deck to where the drone hovered. We were confused. What was he going to do? Throw something at it? Wave to the drone operator through the camera that surveilled us? He came to a stop immediately in front of the winged interloper, his back to us. I looked at Alex's mother, and she shrugged. Bruce had a plan, she just didn't know what it was. This was not new.

Now his shirt was off and his hands had disappeared in front of him, elbows out to his sides. We sat at the table, eyes wide and trained on his back. Could he be fiddling with his belt? Then he was naked, his bare ass toward us. He threaded his hands together behind his head and began to move his hips like a hula dancer. Immediately, off in the distance, we heard laughter from what must have been the drone operators. And just like that, the drone swooped away, never to return.

Bruce crouched, pulled his pants back up, put his shirt back on, turned to face us, and strutted back across the roof like the hero triumphant, accompanied by our roaring laughter. I didn't have an inkling yet, but I would go on to marry Alex eighteen months later. Occasionally people ask us how we met, and I always tell this story. Every time, Alex jokes in response, "One day, I too shall do my family's courting ritual for my son, as my father did before me."

It was moments like this that made me want to stay in Puerto Rico. I seemed to keep meeting the kind of people who, when someone said to them, "You know, you could move to Puerto Rico," thought *You know, I could!* I had made it to Puerto Rico, but staying meant that I needed to square away a few things.

———

Facebook was at that time expecting that eventually there would be a return-to-office date, and thus only people with good performance records could request permanent remote work. I was doing well enough that when I asked my manager what I would need to do to get approved for permanent remote work he said he would talk to HR about getting me an exception. He came back with bad news. Facebook had a policy that it did not permit employees to work from territories, even during company-wide phases of remote work. I could not change my mailing address to Puerto Rico, and technically he was supposed to make sure that I was not going to stay in Puerto Rico much longer. I was already stretching the plausibility that I was just on vacation. When he asked me what I was going to do, I said I didn't know. He encouraged me to not rush to any decision, suggesting I might take a week's vacation and think it over.

I was sitting on weeks of unspent pandemic vacation time. Chris suggested we go out to Culebra, a small island off the east coast of Puerto Rico. Culebra is the kind of tropical island where many people don't bother renting cars because a golf cart is all you need to drive the ten miles from one end of the island to the other. We rented a house on a hill that overlooked the sea and settled down to enjoy ourselves.

While I could have gone anywhere in the United States next, or even most places abroad, in my mind I felt like I had to choose between San Francisco and Puerto Rico. I could stay at Facebook, but I would have to accept the damp chill and loneliness of COVID-19 back in the bay. The idea of returning to San Francisco felt like I would be sacrificing something meaningful about myself, the cold forcing me to spend significant energy in order to ignore my neuropathy that I could have spent living. As I reclined on the porch with my boyfriend, I considered that on the other hand I could choose to stay in Puerto Rico and feel healthier, more like the way I wanted to feel. I would be warm and significantly relieved from the constant tingling in my toes, almost as if my hands and feet had finally been returned to me. I could share a home with two roommates I enjoyed; I didn't have to be alone all day. I didn't want to give up this life.

A few days later I did a video call with my manager and the new manager he was bringing in to run the Inauthentic Behavior pod. I tried not to react as the new manager described his previous background at Facebook. He had been a successful individual contributor on the Growth team, and moving to Central Integrity was his opportunity to become a people manager. It felt a bit like the foxes guarding the henhouse. Growth was a place with clean, straightforward metrics where it was easier to demonstrate that you were a high performer, worthy of promotion.

Combined with the slower promotion path for individual contributors in Integrity, an ex-Growth manager overseeing an Integrity team was a common arrangement, and it caused problems when Integrity came into conflict with Growth. Most Integrity solutions come at the cost of slivers of Growth, as they unwind biases and unintentional consequences that can be introduced by Growth team changes. Slivers matter though when Growth goals each quarter or half might be similarly small. A new Integrity manager with years of experience and deep professional network in the Growth org needed to fight for the concerns of their new safety-focused team, but it seemed implausible they would not feel an urge to compromise too soon when the colleagues they had left behind complained. I felt bad leaving Rahul; he was a once-or-twice-in-a-career level of mentor. Knowing that staying meant losing him anyway took a weight off my shoulders.

"Have you decided what you're going to do?" the new manager asked.

"I'm going to stay in Puerto Rico," I said. "I'm leaving Facebook." What he said next shocked me: "Would you mind not mentioning it to anyone for a few weeks? Tim is also leaving, and we don't want it to seem like a trend." I knew Tim was one of the hardest-working, most dedicated of my peers in the pod. Again, I tried not to react.

271

My original manager, Rahul, shook his head. "You're not supposed to know that," he said. It seemed to me there was something more going on here. Two departures were not a problem — maybe others in the pod were wavering? Maybe if others thought it was a trend, the line might not hold?

"There's no need to rush picking a last day," the new manager added.

People don't pick last days randomly, and exits from tech companies often unfold in predictable patterns. For most tech employees, a sizable portion of their compensation comes from their equity rewards, which is parceled out somewhere between one and four times a year. Therefore in each quarter, the week (or even day) after vesting occurs there's a spike in people leaving companies.

I didn't yet know that everyone in my pod was about to quit or leave Integrity for other parts of Facebook over the same six-week span. Some wanted out badly enough that they bailed just a month before vesting, while others, like Tim and I, waited until right after the vest.

We were coming up on the second vesting date after they had dissolved Civic Integrity. My whole pod had made it past the first vest, which had come in February. I think most of us were willing to give Facebook the benefit of the doubt that it cared about the spirit of Civic Integrity. By the time the second vest had come around in May, people had decided to move on. I got off the call and texted Jeff on Signal, "I told my manager I'm leaving."

"When?" Jeff texted.

"I'm not sure. He told me I could take my time with an exit date. I was thinking maybe in about a month? Give three weeks' notice early next week?"

Within moments my phone was ringing with a call from Jeff. I had not been the best collaborator for the previous couple of months. While I was in San Francisco, I'd been in regular contact with him to answer questions, but since my move to Puerto Rico, it had been substantially harder for him to get hold of me.

"What would you think of me visiting Puerto Rico for a couple of weeks?" he said. He could ask his boss to have the newspaper sponsor his trip. This seemed like a historic chance to document the truth about Facebook.

Jeff flew in a couple of days later. For the first week he was there, I was still trying to wrap up work I had promised my team. I'd invite Jeff over to our condo after work and let him ask as many questions as we could fit in before I had to call it a night at eleven.

Our first day together in Puerto Rico, Jeff and I made a list of themes we each thought were important to cover. Compared to Jeff's, my list was relatively targeted. I think that was because my focus on the subject matter was specialized, a perspective from inside the tech world. I knew I'd need to plan carefully to make a compelling case to the outside world that Facebook knew its algorithms were harming the public and had failed to act. These were incredibly difficult, multidimensional topics. If I was going to make extraordinary claims about algorithms leading to violence, I needed extraordinary proof. Ever since my last year at Google, for the past six years of my corporate career, people hadn't believed me when it mattered, so why would my word be enough now? I made sure to capture photos of any documents needed to substantiate what I would later tell the SEC and Congress.

Jeff's list was broader. He had spent years following what was happening at Facebook. He had a courtside seat for a seemingly endless stream of activists alleging harm from Facebook, only to have their accusations dismissed in corporate communications, press releases, and by talking heads. He had been left with many more questions than answers. I was certain that if Jeff could have kept me awake and functional twenty-four hours a day, he would have.

Each document I read, sometimes at length but often with a quick scan, didn't only paint a richer portrait for Jeff of what was happening at Facebook, it also expanded my own understanding of how Facebook's systems worked. I felt I was giving myself the education I had needed at the beginning of my time at Facebook but never had the time to pursue. My first year at Facebook had been a never-ending crisis without time to even attend orientation. Now I was reading hundreds of white papers explaining how Facebook worked and the known limitations of the system. Facebook, or any other large corporation dependent on algorithmic systems, cannot rely on textbooks or courses at universities to cover business-critical systems for people to use as they develop a minimum level of competency, because they run on closed bespoke codebases. In the absence of a thoughtful resource like that, these organizations would each be more effective if they compiled the hundred most important documents one needed to read in order to get up to speed. It might seem obvious, but often no one identifies and organizes such a collection because the only thing less well rewarded than writing documentation in many technical organizations is organizing documentation.

Now that I had a fuller view into how the Facebook News Feed had evolved over the previous five years, I knew I needed people to understand how a change like Meaningful Social Interactions had failed to be reined in.

Back in 2017 or maybe earlier, the News Feed team had realized that people were posting less and less content on Facebook. They tried lots of different ways to get people to make more content, but nothing worked except giving people more validation from their friends in the form of likes, comments, and reshares. A little hit of internet dopamine goes a long way. Previously, News Feed had been configured to optimize for how much time you spent on Facebook. Now it would be reconfigured to give the most distribution to content that provoked the most interactions. Overnight the content creation rates were turned around. Give people more comments and thumbs-up, and they'll give Facebook more free content to sell ads against.

Yet within six months the Civic Integrity team discovered that new content wasn't exactly free; Facebook just wasn't the one paying for it. Democracy was paying the price. Society was paying the price. Civil discourse was paying the price. In some areas of the world, people were paying the price with their lives. Civic Integrity had sent researchers into Europe in the lead-up to the EU parliamentary elections and heard the same feedback from the right and the left, from big parties and small parties: "We know you changed the algorithm." Researchers for tech companies love it when people make definitive statements like this, because how could anyone on the outside know for sure that Facebook had made a change? Those statements often have the flavor of mythology. They may approximate an individual's perception of the world, but they often reveal more about what the speaker is afraid of than how the technological system itself works. Maybe that fear is loss of control, marginalization, or persecution, but whatever the explanation, it would rarely pass code review by an expert.

The EU political parties were all frustrated with Facebook for the same reason. They had social media experts on their teams who had figured out what content was getting distributed, and it wasn't the content they were most proud of. It was content that elicited anger. Content that made people afraid. Content that brought out the worst in people. There had been a clear *before,* and the new *after.* They didn't know why some algorithm written thousands of miles away determined that this should happen, they just knew that the game had new rules.

Political parties on the right and the left felt torn between running content they knew their supporters genuinely believed in and liked, versus the content that was being distributed. They understood a simple truth: By the time people showed up at the voting booths, Facebook had already voted. You could run on whatever positions you wanted, but Facebook's AIs were shaped by basic optimizations blind to external costs and they would only let the positions that met its business objectives reach your constituents. Just as our home feed prioritization system at Pinterest didn't understand that people got bored, Facebook's algorithms could not see that they were prioritizing certain kinds of content over others, or what the consequences were of that preference. The News Feed prioritization system just knew it could get the clicks.

It had taken a huge amount of effort to shift the company's top-line goal from Time Spent to Meaningful Social Interactions, and it was going to take far more than a few interviews by Civic Integrity staff to shift it yet again. More than a year after those reports came in, the folks on News Feed Integrity were only beginning to build up enough proof to explain what had driven the significant "mix shift" in what was being delivered via the News Feed.

Late in the summer of 2019, a friend of mine who was a core data scientist on News Feed Integrity invited me to a review of his latest research by the senior leaders of the Facebook News Feed. The company had shown tens of thousands of users pieces of content from their own news feeds and asked them to rank how much they liked that content. It turned out that the average Facebook user genuinely liked content from about their twenty closest friends, but that user only genuinely liked reshares from their five closest friends. Put very simply: People don't like reshared content. Fresh original content was seen as part of their friendships. Someone you cared about had taken a moment to create something for you, and people were biased to like it from the twenty people they were closest to. Reshares, on the other hand, were viewed kind of like spam. A reshare requires only a click and you're forcing your friend to see what you think is important.

The conclusion of that review was in its own way a microcosm of how decisions were made and unmade at Facebook. The data scientists explained that they believed a single component of the feed-ranking system caused a significant portion of the problem of disproportionate distribution of extreme content. When Facebook was trying to decide whether to show a user a piece of content, they would predict how likely that user was to like, comment, reshare, or otherwise engage with it. Those were all ways of driving up the total volume of Meaningful Social Interactions that took place on the system. But when that content was reshared, there was a *second* round of interactions that might take place — the friends of the user who created the reshare might like, comment, reshare, or otherwise engage with that content. If Facebook gave greater distribution to content that generated those "downstream" interactions, it would be another way to drive up the total number of Meaningful Social Interactions. To make sure all of that potential engagement was rewarded when Facebook was shaping users' feeds, they added a parameter that they called downstream Meaningful Social Interactions (downstream MSI).

There was only one problem, and that's where my friend's research came in. People won't interact with most reshared content, so Facebook wasn't giving a downstream boost to most posts. Disproportionately, content that really tugged at people's emotion — like fear, disgust for the other, or anger — was able to overcome the bias against reshared content. Downstream MSI might have driven up the total number of interactions on the platform, but it had also provided an express lane for the most extreme content to reach people's news feeds.

Decisions at Facebook are rarely made by an individual (unless that individual is Mark Zuckerberg), so after walking through all the data, the committee of News Feed leaders weighed the merits and came to the conclusion that something would have to be done about downstream MSI. That's where I left the story in 2019 — the most senior leaders for the News Feed said downstream MSI needed to be removed or devalued — and I assumed it had been resolved. But in the process of combing through the documents to explain to Jeff the intersection of algorithms and physical world political implications, I saw they were still trying to fix downstream MSI in April 2020.

Before I'd been taken off Civic Misinformation, I had contributed to the Soft Interventions Working Group, a task force attempting to identify interventions beyond taking down content or people from Facebook (hard actions). After consulting with at least sixty contributors, they had brought a list of ten recommendations all the way up the chain to Mark Zuckerberg himself. Among those recommendations was a big one: remove downstream MSI entirely. When it came to politics, elections, and other civic matters, this would keep the platform from artificially boosting extremism so much.

Removing downstream MSI from "civic" content had happened a month or two before this meeting, and it had been a good first step. But to remove a boost, you first have to know which content is civic. That means you need to write a text-classifier to label civic content using AI. Unfortunately, this was a process that missed, and likely would always miss, the vast majority of what Facebook wanted to catch because of how imprecise language is. To understand that something is "civic" you must understand which topics and issues are political in a society, maybe also who the political movers and shakers are at any given moment. Moreover, only a handful of languages even had those civic classifiers.

Many of the most fragile places in the world are also the most linguistically diverse, and Facebook's business model cannot justify expanding coverage of things like civic classifiers to protect "smaller" languages, at least as these systems are currently built. Training each classifier in each language required thousands if not tens of thousands of examples of civic content, which would need constant updating as new topics moved in and out of being "civic" topics of the moment. Facebook could justify that investment for a language with a billion speakers, but chose not to for languages with less than 100 or 200 million speakers. As an American, I derived an advantage from how relatively linguistically homogeneous the United States is. Other countries are less so. For example, the 2011 Census of India found 14 languages spoken by more than 10 million people. At the time I left, no more than three or four of those had safety systems.

All the same, removing downstream MSI for languages with civic classifiers was a step forward. The Soft Interventions task force wanted to expand that win to all languages, regardless of whether they had a civic classifier, by removing the boost from all content, period. Expanding the removal of downstream MSI in a blanket way had zero impact on revenue, or on the amount of time people spent on the site, or on how often they opened up Facebook. It seemed like a free lunch. Except — *except* — it slightly decreased the number of total MSIs.

I found notes from the meeting that observed: "Mark [Zuckerberg] doesn't think we could go broad, but is open to testing especially in ARC [At Risk Countries]. We wouldn't launch if there was a material trade-off with MSI impact." When Jeff and I looked at that document together, we sat in stunned silence. Here was a simple, content-agnostic intervention. It wasn't about cherry picking which ideas were good or bad, it was just about rolling back a bug Facebook had pushed out — and Zuckerberg was blocking the rollback because he couldn't bear to have any of his numbers go down.

I can't say why Mark Zuckerberg wanted to keep downstream MSI in place, but I do understand the role MSI played for the News Feed team. Goal metrics drive organizational behavior. Having a consistent top-line goal is important because people make decisions about how to invest their time over months or even quarters based on assumptions about how that work will be evaluated. If downstream MSI had been turned off across the board, even just in At Risk Countries (which, in 2020, included the United States), the total volume of MSIs would have gone down. Without declaring a partial amnesty and resetting everyone's metrics, it's almost certain that people would have gotten smaller bonuses if downstream MSI had declined. Teams that had been working on improving components related to downstream MSI would have had their work zeroed out.

There are few people in Silicon Valley who have experience working on algorithmic-recommender systems. Beyond their scarcity, they are in continually high demand because they can drive business outcomes as few other engineers can. At a company like Facebook that was already facing retention challenges, I can guarantee that department heads and team leaders would have been calling Zuckerberg the day after a decision to turn off downstream MSI. They would have raised hell about how it was going to affect their organizations; hurt morale drives churn. I don't know if all of this was going through Mark's mind or not, but I can't think of any other reason why he would choose to protect an arbitrary metric after hearing a team talk about how that metric was needlessly fanning the flames of ethnic violence.

As the number of days I had left at Facebook ticked down, the length of my days with Jeff increased. And they were long. We started meeting in the early morning, locking in a few hours before Menlo Park started for the day. After my workday ended, we would meet again for more work. We were both getting worn out, but as I answered more and more of his questions and I read more and more documents, it was dawning on us how unique and potentially significant this moment was.

Many people have asked me why I was the first to blow the whistle on Facebook with this many documents and covering such a broad set of topics, and I cannot give Jeff Horwitz enough credit. Experts don't always know which knowledge they have is obvious and which is revelatory, nor is it apparent to people inside bubbles what questions might be interesting to those outside. Our knowledge and expertise were extremely complementary in scope. We had started meeting in the woods in Oakland, and here we were. He was helping me see the forest, and I was helping him see the trees. I wanted to act because I knew the stakes were high, but I needed his expertise to understand the narratives Facebook had been telling the public. I could not have collected the information to disprove Facebook's misrepresentations without him.

Over three weeks, we worked from various time-share-style studios that we requisitioned as co-working space. I remember most vividly our days in the last of our studio meeting places. It was located just a few feet above the Condado Lagoon, with a view of the bridge that spans it. To get decent pictures of a computer screen, especially with a 4 megabyte camera, I had to assume a posture no doctor would define as ergonomic. I had tried experiments with tripods for holding the camera, but I found the most efficient way to take clear photos was to place my left elbow on the table that held the laptop, hold the phone in that hand, and then reach down with my right hand to advance the screen before lifting my hand again to tap the picture button on the phone.

After I'd taken at least 15,000 images, my back had begun to hunch even when I wasn't documenting my screen. My pain levels were only kept manageable by the posture-correcting brace Chris had ordered for me. One day, after hours and hours of photographing, I took a moment to space out, staring at the stunningly beautiful blue water through the balcony window. Suddenly a small gray lump briefly broke the surface maybe twenty feet from me. It rose maybe a foot out of the water in a smooth arc before disappearing under the inlet's gentle waves.

"Jeff! Manateeeeeeee!" I shouted, with childlike glee. We had occasionally seen manatees before, but this studio unquestionably had the ultimate manatee-viewing deck. Just then, a second gray object broke the water maybe a hundred feet away in the same characteristic gentle arc, followed less than thirty seconds later by a third one about sixty feet away. "Manatee trifecta!" Jeff yelled. We were getting a little punchy, fueled by exhaustion and too much coffee. Anything that could relieve the tedium and stress of our accountability work was a welcome break.

We pushed into our final week on fumes but were determined to run through the tape. Sunday was our last full session, since my final workday at Facebook would end that Monday at 5 o'clock Pacific Time. That's when I'd be cut off. Jeff and I met outside my condo at 8 that morning and we walked together over to the studio so we could complete our sprint. We had mutually decided that Jeff knew enough about the algorithms of

Facebook and their consequences, so for these last few days we could prioritize topics that Jeff knew were hot-button issues for Facebook even though they were beyond my firsthand experience.

I had decided to take the actions I did because I was worried about ethnic violence, particularly in the most fragile places in the world. But it was largely, if not entirely, because of Jeff's prompting that history has an account of Facebook's impacts on teen mental health, racial and gender bias in AI, advertiser deception, misleading investor-growth reports, and Facebook lying about its preferential treatment of VIPs through its Cross-Check program. I would not have thought to document those topics if Jeff hadn't helped me appreciate what the world wanted to know or, really, *deserved* to know about them.

And Jeff's contribution did not come without cost. By the time Sunday rolled around, Jeff had begun to manifest the stress of this last sprint by regressing in his smoking habit. He had taken to smoking a single cigarette each day, one quarter of a cigarette at a time, in little respites on the balcony when his nerves demanded some fresh air and a chemical backstop.

At some time after one in the morning, now in the wee hours of Monday, I took a break to foam-roll my back and try to stretch out the spasming muscles that had begun to continuously ache, when Jeff returned from his latest micro–smoke break. I'm extremely allergic to smoke, and even the residue of the cigarette smoke on his clothes triggered an allergic reaction. The only problem was that the muscles of my torso were so fatigued from holding me for hours in my unnatural clamshell hunch that I could no longer cough. I sat there hugging a pillow to my stomach, trying to create enough pressure to force out a cough as my body screamed for some resolution to the burning sensation in my lungs.

I can't overstate this. No model of accountability that depends on that level of sacrifice by an individual will be sufficient in the long term for governing a company as powerful and far-reaching as Facebook.

CHAPTER 14

Coming Out

I am not anxious to be the loudest voice or the most popular. But I would like to think that at a crucial moment, I was an effective voice of the voiceless, an effective hope of the hopeless.

—Whitney M. Young, Jr.

The heavy wooden doors parted in front of me and I walked into a United States Senate hearing room as if on autopilot. I was running on maybe four and a half hours of sleep and with each step felt like I was walking through mud, forcing myself forward. If you had sat Jeff Horwitz and me down on my last day at Facebook and proposed so much as the idea, the possibility, that four and a half months later I would be walking into a Senate hearing to testify about what was really going on at the company, that I would be "coming out" publicly as the whistleblower, we would have been horrified. "Absolutely not," I would have said. But now I was sitting at a table, the senators arrayed in front of me, without him.

I would have explained to you that I was working with Jeff because I knew he could tell Facebook's story—he had been one of the leading voices doing this for years. I had captured twenty thousand pages of documents for the public so the case could not be refuted. I was hoping there would be closed-door sessions to address the deeper questions and explain the significance of these documents to governmental officials, but I had no desire to take on a role like that of the Cambridge Analytica whistleblowers. I was a person, not a symbol or a show pony.

I had assumed that I could remain behind the scenes until about mid-July, which is when the *Wall Street Journal* began to provide more concrete information about when they would publish Jeff's stories. When Jeff and I bid each other goodbye in Puerto Rico on May 17, he had told me he would probably publish by the middle of June. That didn't happen. Then the timing got overtaken by other events. The Olympics came along, from July 23 to August 8, and then the summer-vacation season after that. The *Journal* wanted to give Jeff the time he needed, and also to do all they could to ensure that the stories didn't get lost in the news cycle. This meant the first article was going to print after Labor Day. Four full months after I left Facebook.

When that date was firmed up, my lawyers began having serious conversations with me about how I should prepare to protect myself and those closest to me. They asked if I wanted to participate in the rollout, and they began suggesting that perhaps I should, because going public might actually provide more protection than trying to remain anonymous. My lawyers brought in a communications firm that often did issue-advocacy work. At the kickoff meeting, their consultant said he knew someone at *60 Minutes*. He thought if *60 Minutes* could break the story of my identity, they would be willing to do it in a story that covered the revelations. I wasn't yet sold on going public, however.

From that point on, it seemed like at every team meeting the topic would get reintroduced. I would resist, and my lawyers would patiently repeat that they did not think it realistic that my identity could remain a secret. They were clear—Facebook was going to know I was the whistleblower as soon as the *Wall Street Journal* started asking for comment. Facebook would figure out that only one person had accessed the files we used for the first wave of reporting—me. The attorneys argued that I should take responsibility for my actions or I'd spend the rest of my life wondering if this week was when Facebook would introduce me to the world in the least flattering and most destructive way. I should expect Facebook to attempt to destroy me in order to undermine the narrative. Did I want to live with a cloud over my head, always looking over my shoulder?

I began to flirt with the idea. I agreed to an exploratory video call with the producer at *60 Minutes* if Jeff could also participate in it. Already some things were not going to plan, and I was nervous about any more curveballs. During our time together in Puerto Rico, I had been clear with Jeff that I wanted the *Journal* to have the English-language exclusive. After all, Jeff had done the work, he had invested years in reporting on Facebook, and the paper had supported him and taken the risk. But many of the most salient stories in the disclosures dealt with matters that affected people outside the United States. It didn't seem appropriate that an English-language newspaper, behind a paywall in the United States, should be the only outlet to break stories and reveal the truth about how and why people around the world were dying because of Facebook's negligence. Jeff himself admitted that there were many stories within the files that lay outside the scope of what the *Journal* usually reported.

A month before that video call, back in early July when Jeff explained to me the delays occasioned by the Olympics and summer vacation season, I had opined that this was the perfect opportunity to begin organizing a non-English-language consortium of journalism outlets. I hadn't had the energy to get other journalists up to speed immediately after leaving Facebook and moving all my belongings to Puerto Rico, but now I was settled and had time to ensure that we went to press with diverse voices. Jeff wasn't sure about this; it wasn't the kind of thing the *Journal* did very often. He'd need to talk to his editor.

By mid-August, the non-English-language consortium was still not off the ground. As I sat in my home office in San Juan and waited for the *60 Minutes* video call to begin, there were maybe thirty days until the first of Jeff's articles would go live, on September 13. I felt a bit adrift. Less than two weeks after that discussion with Jeff, the guy I had been dating, Chris, slipped in a shower in Las Vegas, alone and under preventable circumstances, and got a frontal lobe bleed that left him irrational and belligerent. Now I was going to keep myself safe. I was staying in the apartment of a San Juan acquaintance, Alex, who was out of the country on business for a few weeks. I was still unsure about going public.

When the video call began and the *60 Minutes* team appeared in various boxes on my computer screen, Jeff and I gave a thirty-minute summary of the disclosures. Then they asked their questions. The next day my communications person told me that *60 Minutes* had greenlit a feature about the Facebook Files, but they couldn't make a collaboration with the *Wall*

Street Journal work. I don't know the details, whether it was corporate politics or an unwillingness to share the spotlight — all I know is that their offer was extended only to me.

Jeff had been my most consistent friend over the previous nine months. Others had provided more support, as I had moved between different COVID-19 housing permutations, but Jeff had been a constant. He had believed me when I felt alone at Facebook. He had given me the support to follow my conscience and been the best collaborator I could have had throughout it all. If I went on *60 Minutes* without him, I was concerned he would feel betrayed. I felt torn. I knew he felt a sense of ownership over this Facebook story, or rather what would be a series of stories, and I knew the *Journal* had been willing to invest in the reporting the way they had, at least in part, because they felt it was a story that they alone would be telling.

I could go on *60 Minutes* and ensure that Facebook couldn't wriggle free from their misdeeds by spinning fables about a disgruntled employee painting a distorted portrait of the company and protect myself in the process, but I would risk alienating Jeff. And what if I fell on my face? What if Facebook dug up dirt on me that they could spin to undermine the *Journal*'s reporting?

In the end I decided to believe my lawyers: the best thing I could do to protect myself was to let the public protect me. And the only way the public could protect me was to let them know me, to come out of the shadows. Just as I believed that Facebook couldn't solve its problems without the help of the public, I didn't think I could solve the likely challenges from the company without the public helping me. I needed to go on *60 Minutes,* and that would risk losing my friendship with Jeff. The *Journal* would launch its series in September with my identity withheld — they alone would have the spotlight — and then I'd come out on national television in October.

With the *Wall Street Journal* a month away from publishing Jeff's first story, I was also a month away from filming the *60 Minutes* segment. I had been asking my communications adviser for weeks whether there was a coordination document that mapped out who was going to talk to whom when the stories launched. I was still committed to bringing non-English-language outlets on board, and I wasn't seeing any movement from my comms team or the *Journal* on making that happen. Nor was there a plan on collaborating further with Congress. If I gave documents to anyone besides Congress or the SEC, I would be facing additional risk to myself. When would we reach out to the non-English outlets and let them know this was coming? When would I give them background briefings so they would know how to navigate the disclosures? Even what I considered really basic necessities were still up in the air. For example, I needed media training so I'd be prepared to answer basic questions on TV like, "How would you describe who you are in thirty seconds or less?" Take a moment—can you summarize your life in thirty seconds? How would I summarize my case into sound bites for a *60 Minutes* audience or a congressional hearing?

The chaos got much worse before it got better. A week or two before that first *60 Minutes* vetting call, in late July or early August, the lead comms person informed me he was partially stepping away from my case but that his lieutenant, a junior employee at the firm, would handle the day-to-day while he supervised. But two weeks before the printing presses started rolling with our story, the senior consultant announced he was off my case completely. I began to panic. I'm no crisis-comms expert, but I knew that only a fool goes on *60 Minutes* without a professional ensuring that they're prepared. I guess in the handoff between the two consultants, details like a rollout plan or speech training fell through the cracks.

I clung to blind faith. Maybe it would be okay. *Oh, please let it be okay.* I willed myself to believe that the junior comms person could get me ready. I met with him the next day, and to my horror discovered I was starting from scratch teaching him about the issues in the tech accountability space. He wasn't going to be able to help me sharpen my answers if he wasn't already familiar with the ideas we were talking about. We met several times over the next couple of days before I realized I had to find outside help.

If you watched my *60 Minutes* appearance or my session before the Senate two days later, all of this might come as a shock to you. My delivery was smooth and my demeanor calm enough that it had inspired a cottage industry of conspiracy theories featuring crisis actors and shadowy billion-aires behind the curtain. At this point Larry Lessig stepped in and provided me with critical support and counsel. Our paths had crossed briefly that summer when a friend introduced us, and Larry had told me to reach out if I needed help. A Harvard Law School professor and the former director of The Edmond & Lily Safra Center for Ethics at Harvard University, Larry is the founder of Creative Commons and has long been an advocate for fairness and transparency in tech. He steered me toward the communications firm Bryson Gillette. They were a natural choice for a last-minute save because they had been supporting the Center for Humane Technology for two years, during the release of their documentary *The Social Dilemma,* which dug into issues similar to my disclosures. Bryson Gillette didn't need hours upon hours to catch up on background information. Of all the communications firms in the United States who could have gotten me up to speed, they were almost certainly the best choice. As my mother used to say when I was small, I have a very active guardian angel.

My prep schedule was intense over the six days before the *60 Minutes* taping. Saturday brought the first ninety-minute session with Bryson Gillette, followed the next day by four hours of answering hard questions with my lawyers. Monday I was back with Bryson Gillette again for an hour when the *Journal* published the first of Jeff's articles, "Facebook Says Its Rules Apply to All. Company Documents Reveal a Secret Elite That's Exempt." Tuesday was what I considered my dress rehearsal. I recorded a podcast with the *Journal* that would be released simultaneously with the profile the *Journal* would publish when *60 Minutes* aired. We knew it would be a friendly interview and they would give me a safe space in which to practice explaining the core issues of my disclosures and describing what I had done.

Sometime in the week or two preceding the release of the Facebook Files, my lawyers reached out to the Senate Commerce Committee's Consumer Protection Committee to brief them on the upcoming news. This committee had been investigating social media for months, and they jumped at the idea of my testifying after I came out. I had never talked to anyone from the committee until the week the stories began their rollout, but sometime during my packed schedule in D.C. I also met with two of the committee's staffers. One represented the majority chair Richard Blumenthal's office; the other was from the minority chair Marsha Blackburn's office. This meeting felt something like an interrogation, with the two of them seated across from my lawyer and me at a small conference room table.

Their mission was simple: Were my disclosures enough to warrant a Senate hearing? Would I make a good witness? Did I sound credible? Competent? Their hours of questions gave me a sense of what issues we'd want to tee up during the *60 Minutes* taping. The hearing was soon scheduled for two days after the show would air. This session was also the first time I realized that there was a difference between what motivated me to come forward and what the public would be most interested in. I came out because I thought there were tens of millions of lives on the line in some of the most vulnerable places in the world. The staffers were most interested in the mental health of teenage girls and what Facebook knew about it.

Just six weeks earlier, at the start of August, the Commerce Committee had drawn attention in a public letter to a few comments Mark Zuckerberg had made when he had been hauled in front of Congress in March of 2021 to explain Facebook's response to the events of January 6. He had stated he believed Facebook was conducting research on the effect of its products on children's mental health and well-being. The committee's public request was clear: "Has Facebook's research ever found that its platforms and products can have a negative effect on children's and teens' mental health or well-being, such as increased suicidal thoughts, heightened anxiety, unhealthy usage patterns, negative self-image, or other indications of lower well-being?" The senators demanded to see any research Facebook had concerning children's mental health.

Two weeks later, when Facebook's response was due for release to the Senate, Jeff had gotten his hands on a copy of the company's statement in response and called me to get my thoughts. My eyes widened as we read through it together, amazed at Facebook's hubris. Facebook had a pile of research interviews with children who said they compulsively used Instagram even though it made them unhappy, interviews their own researchers summarized as being "addicts' narratives": *It makes me unhappy; I can't stop using it; if I leave, I'll be ostracized.* And yet Facebook's response to Congress was little more than a shrug: "We are not aware of a consensus among studies or experts about how much screen time is 'too much,'" they responded, spinelessly sidestepping the actual question asked, before producing eight hundred words detailing various media literacy programs they had funded and user experience controls they had built.

Facebook was technically correct that there was no consensus in academia or the public about what amount of social media was "too much." But that was in no small part because Facebook kept all of its data hidden and wouldn't let academics have access to it for studies. Yet. Congress hadn't asked how much social media was a "bad" amount of social media; they had asked whether Facebook's research ever indicated that kids might be harmed by Facebook's products . . . and now I was sitting in front of congressional staffers who were thumbing through printouts of that very research. They were not pleased.

Instagram is probably a positive force in many kids' lives, but like all harms on social media, the negative impacts of the platform are not evenly distributed. For any given type of harm caused by the platform, whether we're talking about kids or adults, you should expect around 10 percent of users to bear an outsized amount of the burden, because that small fraction comprises by far the heaviest users. Just like with the 90-9-1 Rule for Participation Inequality in Social Media, where a small number of users produces almost all the content, when it comes to *consuming* content on Facebook, we see a similar dramatic skew. The top 1 percent of adult users might consume two thousand to three thousand posts per day, while the average user consumes only twenty or thirty. In the case of "problematic use" — Facebook's phrase for Facebook addiction — Facebook researchers found that kids' problematic use rates rose from 5 percent when they were fourteen years old to 8 percent when they were sixteen.

That might not sound too bad, but in 2018, when the study was done, Facebook had long since ceased being the most popular social media platform for teenagers. Instagram and Snapchat held that distinction. Also, the study's definition of "problematic use" was extremely stringent. It didn't just mean the user had "serious problems with your sleep, work, or relationships that you attributed to Facebook"—the user also had to have the self-awareness and honesty to admit that they also had "preoccupations" about how they used Facebook, like having a fear of missing out or lack of control. If 8 percent of sixteen-year-olds could admit that they had "problematic use" on Facebook, imagine how many more might have admitted problems with Instagram, let alone the ones who would have denied it despite struggling. And that was just compulsive use. Other harm types were concentrated as well. Thirty-two percent of teenage girls said that when they felt bad about their bodies, Instagram made them feel worse. Other internal documents talked about self-harm communities that encouraged kids and adults to hurt themselves.

In my mind, the critical question wasn't whether Instagram was safe or not, or safe enough for enough kids, as important as those questions were. It was whether a company that felt it needed to hide its own research from the public should be the one to decide when "too many" children had been harmed.

Facebook's own research suggested that not all social media platforms were perceived by kids as equally harmful, illustrating how important design and experience choices are when it comes to the consequences of social media. TikTok was associated with playfulness, performance, and doing fun things with your friends and even family. Snapchat was primarily a chat app. When it focused on the body at all, it didn't go beyond the face, which you could augment with playful virtual masks. But Instagram was about idealized lives and bodies, and many teens felt they didn't measure up. Instagram was designed for social comparison.

Safely using Instagram required a sophisticated level of self-awareness that even adults struggled with. You had to understand that the lives people shared on Instagram were not representative of their actual lives—that people only post the most carefully curated best-life moments. Instagram tried to respond by pushing out marketing initiatives about being your "real self" on the platform. Yet marketing dollars, even at Facebook scale, can't fight the tide of adolescent desire to be accepted and respected by their peers. The result was that, especially for young girls, who often were not even old enough to be on the platform officially, Instagram fueled negative social comparisons that left them feeling like they didn't stack up against their peers.

The American Psychological Association explains in an article titled "Why young brains are especially vulnerable to social media" that part of why social media has such an impact on children as they go through puberty is that children's brains evolve to have more dopamine and oxytocin receptors — social-reward neurotransmitters. As a result, any adult reading this will literally never get a compliment as sublime or a criticism as painful as what a twelve-year-old girl may receive. Most adults reading this would think *Thank God.*

If you think the hallway of a middle school is a tough place to be, try social media. When those schoolyard bullies, taunts, and cliques move online, the child on the other end of an insult becomes more abstract and it's therefore easier for one immature child to hurt another. A mean zinger gives the sender the satisfaction of lashing out, without the negative feedback of watching someone cry. Insults and compliments persist online indefinitely, making concrete and permanent the pain that was once ephemeral.

Part of what the congressional staffers I met with found most disturbing was how Facebook employees referred to children who used the company's products. Facebook referred to ten- through thirteen-year-olds (tweens) as "herd animals." Internal marketing studies flagged protective older siblings teaching middle schoolers how to safely use Instagram as a "problem," not because they were bringing in users younger than the thirteen-year-olds allowed by Facebook, but rather because such siblings would coach their younger brothers and sisters to be judicious about sharing personal details. They would (rightly) tell their siblings that once you posted, your content might be on the internet forever. In the eyes of Instagram growth-hackers, these were not caring siblings, these were dangerous people spreading what they described as "myths" that might create, according to the wording of the internal documents, "sharing barriers for upcoming generations including that being spontaneous/authentic doesn't belong on Instagram."

Then came the algorithms. Facebook had been blunt about blaming consumers for what Facebook's Recommender systems provided. In March 2021, Nick Clegg, president of the company's Global Affairs division, went so far as to write a blog post titled "It Takes Two to Tango." In it, he attempted to refute the criticism that Facebook's algorithms were radicalizing people and pulling them down rabbit holes by pointing the finger right back at the user. Facebook's algorithms relied on who your friends and interests were, Clegg wrote, and both of those were things that users chose, not things pushed by the Facebook algorithms. "Don't wag your finger at us," he seemed to be saying.

But Facebook's own internal documents showed that Facebook and Instagram algorithms consistently promoted and pushed more and more extreme content over time as they blindly chased clicks and comments to drive up their content-agnostic goal metrics. The clearest real-world example of this I've heard, I got from a reporter. The reporter had a brand-new healthy baby boy; he created a private Instagram account dedicated to the baby to share his joy with his friends and family. The baby was healthy and cute, and the account only had perhaps five other friends, all of them other cute, healthy babies. These accounts only posted sweet baby pictures. And yet, about 10 percent of the reporter's feed was mangled babies. Some were visibly misshapen, some were in the hospital with tubes coming out of them, some had wounds. Many exhibited observable distress. This parent never wanted and never sought out images of suffering babies. How did we get from happy, cute babies to a father's worst nightmare? The answer: One of the signals in Instagram's algorithm was almost certainly dwell time, which is the amount of time you leave an image up on your screen. It didn't matter that the reporter hadn't clicked on the upsetting images or left a comment about them. Instagram learned that people interested in babies will be unable to mindlessly scroll past a suffering child, so you'd dwell on horrifying images you most definitely never wanted.

When it came to teenagers, the topics escalator wasn't from chubby cheeks to tears, but rather from innocuous topics like #healthy_recipes to pro-anorexia topics. Or from #extreme_dieting to self-harm. All pushed forward by the same blind algorithmic escalation. Child psychologists have since told me how hard it is to care for a child struggling with an eating disorder who tells you they are being followed around Instagram by their past unhelpful choices.

One question the congressional staffers asked me stands out in my mind. I've spent hundreds of hours in the past year working with journalists, government staffers, and nonprofit groups helping them understand the content of the disclosures, and part of why that time is so worthwhile for me is that I always learn something from the questions I'm asked or the lens other people bring to the documents. I found my time with these staffers incredibly educational. One of them pulled a folder of printouts from his stack, flipped to the third page, and said, "There's no way this can really mean what it means, right?" The document featured the pitiful title "Will teens eventually adopt Facebook?" Under the words was a banner image of a sloth wearing a marathon bib sprawled across some pavement, one claw outstretched to cross the finish line. It was dated 2018. "Is this proof they've known for years they have lots of users under the age of thirteen?"

I stared at the graph. Along the bottom it read "Approximate Age" and along the left it read "Monthly Active People/US Population" (basically, the percentage of the population active on Facebook). The graph itself was a number of overlapping S-shaped product-adoption lines, one for each birth-year cohort. The story the graph told was clear: It used to be that teenagers joined Facebook very rapidly. Now that Facebook was "uncool," teenagers joined much more slowly. Of the children born in 2004, only about 30 percent of fourteen-year-olds (in 2018) had accounts. For children born in 1997/8, 80 percent had accounts. This wouldn't have been shocking, except the lines extended back to what fraction of their cohort was online when they were ten years old.

The staffers asked whether Facebook *knew* that ten-year-olds were signing up for Facebook. I, too, was confused. Facebook's weak attempt to say they cared about keeping children under the age of thirteen off the platform was that they asked for your birth date when you signed up and blocked you if you said you were under thirteen. Of course, you could lie about your age. But there was no way to be "ten years old" on Facebook, officially.

Then it leapt out at me. The graph didn't say "age," it said "approximate age." This was a big deal. Every platform that makes money through advertising has an estimated age for each user, because it makes the platform far more valuable to advertisers. Children aren't the only ones who lie about their ages when they sign up for a new account; adults do, too. Advertisers still want to target you with ads for a fortieth birthday tropical getaway vacation package.

If I hadn't spent those hours with the staffers, I would never have known about this graph. People often ask me questions like "What can we do to keep children off of social media?" If the most dangerous window for children on social media is under thirteen, do we have to follow China's lead and start tying our driver's licenses and/or Social Security cards to our social media accounts to keep our children safe?

Facebook has many ways of figuring out how old children are. Some of them are simple: some children write in their Instagram bios that they're a fourth grader. Most users have location tracking on. (Take it from me: turn it off!) If you show up at an elementary school every day and most of your friends aren't adults and show up at the same elementary school, you're probably a child. Children report each other to Instagram when they want to punish each other: You snubbed me at recess, now you're losing your account. We've known for ten-plus years that who you socialize with online and how you behave exposes tremendous amounts of information about you. Combine that with the topics you engage with or how you communicate, and it only takes Facebook finding a few thousand children, let alone ten thousand children, for them to build machine-learning models that can easily find functionally all of the under-thirteen-year-olds in a country.

The reason why Facebook doesn't do this today is they know getting kids on their platform as early as possible is the key to having a platform in the future. All of the social media platforms know children age with their platforms. There's a reason why people joke about Facebook being for millennials — it was their social media when they were twenty-one years old, and it still is when they're thirty-five. All the platforms are locked in a standoff — none feel they can be the first mover to protect kids, because the risk of losing the next generation is too high.

Let's imagine that Facebook had to share with the public that graph the staffer had dug out of the Facebook files, updated every quarter, showing the full and complete number of users they thought were fourteen. How many of the platform's current fourteen-year-olds had been on the platform already, for two, three, four or more years? Suddenly there would be a yardstick to compare how much each platform was actually doing to protect children. It would take Facebook less than a day, and only once, to produce a pipeline that would provide that ongoing report. They don't share it with the public because they don't want the public to know the true extent of underage children on the platform.

It is this basic level of transparency that many jurisdictions around the world—the EU, UK, Australia, and Canada—are finally demanding of companies such as Facebook. Just because platforms don't share data today doesn't mean we couldn't live in a world in the near future where they were required to publish at least the minimum amount of data we need to assess whether they're fighting hard for our safety. Then we could know that they're resolving conflicts of interest behind closed doors the way the public would want them resolved.

As my time with the congressional staffers drew to a close, I asked them if there were any other committee members we should give the documents to. After all, there were thirty senators on the committee, and each must have at least one staffer covering tech-related issues. They looked at one another and then turned back to me. "I know that sounds really reasonable, and I would say yes [the staffer paused, conveying how little he wanted to finish his sentence] if we could go two weeks without a committee staffer leaving for a job at a Big Tech firm."

I stared at them for a moment. This, too, is what the public is up against. Just as an engineer can't take even one college class focused on how Facebook or any other social media platform works and the interplay between the algorithms and the product choices, none of the staffers on Capitol Hill had been able to take an analogous course when they were studying political science or public policy. They were learning on the job about tech regulation—which safeguards were available to protect the public, and which ones were available to protect the tech companies—and as soon as they reached a certain level of fluency, they'd be poached by Big Tech with salaries five times what their senators could afford to pay them.

The morning after that meeting, I woke around seven and took the elevator down to the basement of the hotel where CBS had put me up. I looked around sheepishly, uncertain where to go in the cavernous convention center hidden beneath ground level, until I ran into someone exiting the ballroom. That person led me into the dimly lit space. I was startled to find what looked like a film set waiting for me. Bright lights surrounded two chairs that faced each other, small pieces of tape on the floor marking their positions exactly. I stood there surveying the scene, not knowing what to do next. The closest thing to a film set I had ever been to before was in the studio audience of a live episode of Bill Maher's show ten years earlier. I never expected to be the one sitting in a chair like these.

An assistant spotted me hovering and walked over. "Scott's not here yet. Why don't you get your hair and makeup done, and he can go next?" Scott, as in Scott Pelley, the silver-haired, baritone *60 Minutes* reporter. I sat down with the makeup artist and she blew out my freshly washed hair. Twenty minutes after that, for the first time in my life, I had a fully made-up face.

I remember taking a picture of myself and sending it to my new comms person. "I feel like I look like a Texan," I tapped into the phone. My blond hair was straight, with a volume I didn't think possible, conjured by a blow dryer and hair spray. I didn't feel like myself. This is why I never wanted a role as a whistleblower. Was this what my future was going to be? Jeff had come out to D.C. for the week, in part so we could have dinner and celebrate the day the first article in the series was published. On the set, viewing myself in a mirror, I thought back on the advice Jeff had given me over Thai food. "Remember, you don't owe anyone anything. You can stop at any time. It can be a week from now, a month from now, a year from now. It's your life, and no one else's." I had already done what was needed; anything above that was a choice.

I surveyed the scene and the film crew that surrounded me. I felt like I'd already tied myself to the front of the train. I had made my decision. The momentum of the filming would carry me through to whatever came next.

I was told the crew would need an hour or at most ninety minutes to film. My segment would have a run time of about thirteen minutes. Two and a half hours later they called a wrap. I had been warned that Scott Pelley might ask the same question multiple times in slightly different ways, but that hadn't happened. He told me afterward our conversation was fascinating, and he could have gone on for another hour. I didn't know this yet, but this would become a trend. The world was hungry for a look behind the curtain.

My comms team had sent me a couple of examples of *60 Minutes* interviews for me to watch beforehand, and it made me more aware of the ways people deal with being nervous and the risks of being featured in a thirteen-minute segment composed of snippets pulled from an hour or longer interview. I maintained eye contact with Scott almost continuously for two and a half hours and held my shoulders back to make sure my posture never slumped because I'd seen that any moment could be snipped out and included if the sound bite was juicy enough. You don't think about it in your daily life, but it's amazing how much slumping erodes your trustworthiness. The cramps in my upper back reminded me that credibility has its costs.

I flew home to Puerto Rico the next day. The fires in California had been so bad that year that smoke had extended all the way east to D.C., and the sunsets every night in the nation's capital evoked the end of days. I did my best not to interpret it as a personal omen. Stepping out of the San Juan airport, I took a deep breath of the fresh tropical air, my lungs and body able to fully relax for the first time in a week. I had filmed my debut for the world.

Jeff and I still didn't know exactly what the rollout schedule would be for his series the week before the first article graced the pages of the *Wall Street Journal*. We thought there would be two or three stories a week, spread out over time. Instead the *Journal* went with what we referred to as Shark Week, after the eponymous week of shark documentaries that once was a fixture on the Discovery Channel. Every day for a week, a new installment of his long-form journalism dropped, adding up to over 20,000 words in all. It was a lot. And the world noticed.

Monday detailed Facebook's hypocrisy of insisting all users are treated the same while letting celebrities run free. Tuesday brought "Facebook Knows Instagram Is Toxic for Teen Girls — Its own in-depth research shows a significant teen mental-health issue that Facebook plays down in public." This flowed into "Facebook Tried to Make Its Platform a Healthier Place: It Got Angrier Instead — Internal memos show how a big 2018 change rewarded outrage and that CEO Mark Zuckerberg resisted proposed fixes" on Wednesday. Thursday took a more international focus with "Facebook Employees Flag Drug Cartels and Human Traffickers. The Company's Response Is Weak, Documents Show — Employees raised alarms about how the site is used in developing countries, where its user base is already huge and expanding." And Friday tied up a number of threads woven through the previous days with "How Facebook Hobbled Mark Zuckerberg's Bid to Get America Vaccinated — Company documents show anti-vaccine activists undermined the CEO's ambition to support the rollout by flooding the site and using Facebook's own tools to sow doubt about the Covid-19 vaccine." It was an almost perfect case study of how different flaws in the product built on each other until 4 percent of users received 80 percent of the COVID-19 misinformation. Part of why otherwise reasonable, level-headed people showed up and screamed at school board meetings and sent death threats to teachers was because their feeds were swamped with people making them fear for their children's safety.

Then nothing for ten days. My lawyers and I were confused; we knew there were more stories, some of them particularly shocking. What could be going on inside the *Wall Street Journal*? My *60 Minutes* segment was to air on October 4. I knew I was going to spoil some of the news because I had talked about it during my interview. I thought Jeff would have published the whole first wave of stories by the time it went on air. It was out of my hands now. All I could do was to wait.

Many of my friends had known for months that I was a whistleblower for Facebook, and they kept asking me if I was going to watch myself when *60 Minutes* aired. Each time I would say, "If I can help it, definitely not." Part of how I had survived the chaos of the first few weeks after the articles came out was by not reading press coverage of them. If you don't look, you don't have to see how it's received. Now I was about to step out onto the world stage—a single person standing up to a trillion-dollar corporation and accusing it of wrongdoing that put hundreds of thousands, if not millions, of lives on the line. Very soon I wouldn't be able to ignore how people responded.

Two weeks later I flew back to D.C. I had one day left of anonymity. My congressional testimony was scheduled for October 5, the Tuesday following the broadcast. While Bryson Gillette had stepped in to provide communications training, I still had a terrifying amount of communications preparation to work through myself. My friend Michael had come to Washington to support me during my testimony (an offer I initially resisted, and now am endlessly grateful for), and he set about helping me get through that critical backlog of preparations.

One of those tasks was creating a web presence for me. Some critics pointed to the "slick professional website" that we put up as some kind of proof that I was out for personal gain and actively seeking the spotlight. In fact, with help from others, we just barely got that website up on Squarespace, a basic website provider, at 1 a.m. the morning before *60 Minutes* aired. While we threw the site together, we drank cup after cup of chamomile tea trying to manage our stress. It wasn't copyedited until hours before the broadcast went live.

Social media accounts were quickly secured, whistleblower-specific Twitter handles and domains registered, email accounts for media inquiries set up. The only headshot I had for six months was taken outside my lawyers' office the next morning. Among the rapidly snapped images, the best one was touched up with the help of Michael's sister's finesse with Photoshop.

The next evening, Sunday, after a long day of mock Senate testimony that left me exhausted and everyone else in the room nervous, Michael and I took scooters across D.C. to have dinner with a few mutual friends who were in town for the week. Senate prep and our commute time by scooter had both run longer than we had expected, and to my relief by the time we arrived my *60 Minutes* segment had already aired. Everyone else present had seen it live, and we spent twenty minutes eating takeout Thai food (a reliable gluten-free option) as they animatedly discussed it. I was exhausted after thinking on my feet and integrating critical feedback all day. I was happy to let them talk about it if it meant I didn't have to watch it.

At some point they insisted they put it on for a second showing. I sat there petrified, holding a plastic takeout container of soup I didn't touch, as the recording rolled. My Texas hair was still there, and it still felt like I was watching a simulation of myself, but it wasn't the nightmare I'd feared, it was...okay. It seemed that the producers really understood the documents and had digested the additional context we had walked them through. They had made an incredibly complex topic understandable by a general audience, and I sounded as if I knew what I was talking about. Between those thirteen minutes of television and the profile of me the *Journal* had published on their website when *60 Minutes* had started running teaser trailers with my name, it seemed about as good an introduction to the world as I could have hoped for. We stayed for maybe another half an hour before heading back to our Airbnb to recharge.

We awoke Monday to another day of meetings, ranging from more mock-Senate testimony to spending two hours answering questions from congressional staffers. People have often asked me why my Senate hearing was so different from Mark Zuckerberg's infamous one. He had appeared before a joint session of the Senate Commerce and Judiciary Committees three years earlier, in 2018. That was the session in which Senator Orrin Hatch asked, "How do you sustain a business model in which users don't pay for your service?" and Mark, after staring him down almost as if he was trying to decide whether Hatch was serious, calmly replied, "Senator, we run ads."

Part of why my appearance was different is that the world had evolved since 2018. But a second factor was that a lot of staffers had worked hard to make sure their senators had the information and tools they needed to be prepared for the session. By the time my *60 Minutes* episode aired, the *Wall Street Journal* had run multiple long pieces in their Facebook Files series, their reporting providing a primer tens of thousands of words long. Then, on that Monday, more than twenty staffers dialed into a Zoom call to ask questions to clarify their understanding of what they'd read. Before Mark's grilling in front of Congress, none of that preparation had been available.

None of the corporate information they needed to know was public. Facebook had been able to define and frame the conversation however it best suited them.

We paused the Senate prep that afternoon for lunch and a briefing by the security consultant my lawyers had hired. Unsurprisingly, there was chatter in some of the darker corners of the internet about me in the wake of the reporting, but the breadth of the focus shocked me. Michael and I had stayed up most of the night hunting the internet for mentions of me and taking my accounts dark, but we hadn't thought to do the same with my mother. Because the army of conspiracy theorists couldn't directly scrutinize me, they scrutinized her. The security consultant told us the good news was that she hadn't found any death threats yet despite combing 8chan and the darknets.

I had started dating Alex the previous week and messaged him on Signal chat from my laptop, "Hey, we haven't been dating very long. I'm sitting in a security briefing listening to a threat researcher tell me what the darkwebs and 8chan are saying about my mother. You don't know what's coming. If this is ever too much—I won't judge you if you need to pull the ripcord and bounce."

"I'm not afraid of trolls," he responded, "I am a troll." I would learn only later that he had a collection of troll accounts on Twitter he enjoyed provoking Chinese information operations with. Everyone needs a hobby.

One of the things people fear about being whistleblowers is that it will end their life as they know it. I've had countless people start thank-yous by saying, "I'm sure this destroyed your life, but thank you . . ." Before I came forward, I certainly feared the same thing. This was the first time I thought, "Maybe this will be okay in the end."

October 4 was a particularly special day for a few reasons. I had taken up Sean McCabe's name as a totem of protection in the aftermath of his death because he loved being the center of attention, and he was fearless. October 4 was my first full day out in the world with my own name, and it was also Sean's birthday. He had loaned me his name for most of a year, and on his birthday I returned it to him with much gratitude. I could now be openly myself.

The second reason October 4 was a special day is that Facebook went down for the second-longest outage it had experienced in ten years. Around 11 a.m., as we sat around the long conference table rehearsing answers to hostile questions (none of which ever came), someone announced. "Facebook is down!" Facebook, Instagram, Whatsapp, and even the keycard readers that grant access into Facebook's data centers had all disappeared from the internet in the blink of an eye. Conspiracies ran amok — people kept forwarding tweets reading, "Frances took down Facebook!" The outage highlighted how much power was held in so few hands and what the consequences were of not requiring a certain level of quality. The next day, Facebook reported that the entire outage had been caused by one person pushing a flawed command out to the backbone network that connected their data centers, and a bug in the auditing software that was supposed to prevent actions that could be so cataclysmic to the global communication infrastructure.

We ended our prep for my Senate testimony around five, and they released me to relax before my big day. Unfortunately, sometimes relaxation isn't under your control. The chaos began around eight that evening as Michael, Inga (another friend who would be supporting me the next morning), and I were walking home from dinner. My phone lit up with a call from Jeff. He was panicking over the fact that 60 Minutes had posted online summaries, including excerpts, of the documents they had used in their reporting. Those summaries were scooping reporting Jeff had ready to go but hadn't yet published. I called the 60 Minutes producer we worked with, and she said my lawyer (not Andrew Bakaj) had told her she could publish them. I called the head of the legal aid organization that was helping me and told her what 60 Minutes had claimed; she promised to get to the bottom of it and hung up.

My heart was racing. I already felt guilty for complicating the Journal's reporting. Now my interview had inadvertently undermined their rollout. Jeff was unsure if the stories he had written would even run because people were already pushing out quick-hit stories covering what 60 Minutes had posted. It wasn't news anymore. In the end, all I could think is that somebody, somewhere, among my lawyers and the 60 Minutes team, had screwed up. Jeff blamed my lawyers. My worst fears seemed to be becoming real.

Part of why I advocate so strongly for independent nonprofits that support whistleblowers is that we need third parties who can keep tabs on individual whistleblower law firms and help direct clients to those who will best serve them. I had nearly gone on *60 Minutes* without communications training because I made an understandably naive choice. Now I was worried I had alienated my closest collaborator because of sloppy communication in the heat of the moment. Given my exhaustion and the rate of change in my life, hearing Jeff's distress over whatever had taken place between my lawyers and *60 Minutes* pushed me over the edge. I sat on a curb in front of a brownstone and cried.

Inga sat next to me and rubbed my back until I regained equilibrium. We walked home to our Airbnb and once again heated water for chamomile tea. But I couldn't sleep. The adrenaline was still running through my veins, mixing with pre-Senate jitters. By the time I walked into the hearing the next day I was at the point where I was just pushing through. I had only one goal — survive the next two hours.

The Senate hearing room inside the Russell Building was a long rectangle, and we entered through a side door. Running the length of the chamber was a long U-shaped dais where the senators sat. The senators seemed to loom above me, but I've since looked at pictures of that day, and the floor is level. I guess it says something about the intimidation I felt when they eventually took their seats. Only you can make yourself feel small. I walked over to the witness table and sat down. Out of nowhere press photographers appeared and swarmed around me for about two minutes. Once the hearing began, they wouldn't be allowed to get between the senators and me, and they were making the most of their remaining minutes of access.

The committee was called to session and went by in a blur. The chairs read opening statements, and then I read mine. After the obligatory opening statements, I launched into the heart of the matter: "My name is Frances Haugen. I used to work at Facebook and joined because I think Facebook has the potential to bring out the best in us. But I am here today because I believe that Facebook's products harm children, stoke division, weaken our democracy, and much more."

I spoke for just a few minutes. Each committee member took their five

minutes to ask questions, sometimes (thankfully) veering off into a monologue for minutes at a time, which gave me a chance to regroup. The questions were largely good and relevant. If the staffers had seemed upset that
Facebook had lied by omission about the impacts of their products on children, the senators gave the impression that they were even angrier.

Silicon Valley has a certain hubris, a conviction that it is a world unto
itself. That Big Tech companies are their own nation-states with embassies,
not offices, in major cities around the world. There is a reason why Facebook had spent hundreds of millions of dollars convincing the public and
regulators that there was nothing that could be done to make social media
safer, all the while spending the majority of their safety budget in the
United States though only 9 percent of their users lived there. That reason
is that Congress actually governs Facebook. Despite what Mark Zuckerberg may believe when he's on the West Coast, Facebook is *not* a nation-
state unto itself.

Woven through my answers was my attempt to lay out the argument
as cleanly as possible: Facebook was full of kind, smart, conscientious people, who were limited in how they could act by a system of governance that
showed no signs of changing on its own. Facebook had built a corporate
culture that valued "objective" measures like computable metrics over
human judgment. This empowered their youngest employees, who fueled
the social media innovation that the company required. But the philosophy
consistently missed preventable problems until too late. It's hard to measure the impact of stopping a fire before it rages, and corporate cultures too
wed to navigating by "objective" numbers instead of human judgment are
at risk of creating "arsonist firefighters." Facebook's problems had grown so
severe because they knew they controlled all the cards. Its closed software
ran on data centers no one could inspect. As a result, it could deny, deflect,
or minimize any concern brought to its leaders about their products. Facebook will not change as long as the incentives do not change — as long as
the only metrics they have to report externally are about the economics of
the business, they will keep paying for those profits with our safety. These
problems are not isolated to Facebook. I had worked at other algorithmic
companies, and understood why sensible people at Facebook had made the
choices they'd made. The same challenges exist in any place where the

public lacks transparency to ensure conflicts of interest between safety and profits are resolved in the public's best interest. Hope and change are possible. We can demand transparency and the social pressure that comes with it to incentivize changing algorithm design and the design of social platforms themselves. Facebook claims we must choose between the status quo and extreme censorship. This is not true. Facebook knows how to solve its problems without picking "good" or "bad" ideas, it just feels it can't be the first mover to use those interventions or the stock price would plummet.

About two hours in, I thought I was reaching the end because all the senators in the room had taken their chances to speak, but then new senators kept cycling in. All ordeals are easier to endure if they have finite ends. I felt like Alice in Wonderland, a place where time didn't matter. The hearing went on for three and a half hours as I patiently answered question after question, ranging from children's mental health to terrorism to interface design to advertising policies.

At last they thanked me for my service and let me go. I returned to the waiting room where they'd stationed me before the hearing and tried to eat a small snack. My throat was sore and my voice a little hoarse, but I was done.

I stayed in Washington for three more days, returning to the Capitol multiple times for smaller private meetings with groups of senators or House members. Perhaps the most interesting one was with a group of Republican senators. They were skeptical that anything could be done. In their minds this was a tragic conflict between freedom of speech and safety. Yes, the reporting was harrowing. Yes, Facebook had demonstrated they could not be trusted. But censorship was a bridge too far.

Their fatalism was understandable. Facebook has spent hundreds of millions of dollars spreading the idea that the only way we can make social media safer is by censoring it. They invested $280 million alone in the Facebook Oversight Board—a body that sounds like it was created to oversee Facebook as a whole but in reality only made decisions about whether individual pieces of content were appropriately moderated. In 2022 alone, *Politico* reported, Facebook spent $20 million lobbying the federal government; they had the largest lobbying operation in the entire tech sector. When you read a statistic like that, you almost must ask, "*Why do you need such a large lobbying organization?*"

I explained to the senators that there were many ways to make social media safer. Specifically, and it seems to me most obviously, social media can be designed to be safe from the start, instead of tacking on content censorship systems *after* the algorithms and design choices that prioritize the most extreme and divisive content.

Some changes were simple. If the senators cared about personal responsibility, Facebook should be respecting the choices users make about what content they want and do not want to receive. The company's data scientists had found that Facebook regularly showed content it thought you "might like" over content you had asked for from your family and friends, or groups or pages you had actually joined. Facebook had run experiments where greater priority was given to content you had actually chosen, and for free, you received less violence, less hate speech, and less nudity. Contrary to Clegg's disingenuous tango analogy, your friends and family were not and are not the problem. The real problems were hyper-viral algorithms and recommender systems blindly optimizing their business metrics at the expense of public safety. But it turned out more content from your family and friends was a little less riveting as well, and you spent a little less time online, something that might send the stock price into freefall if it were to find its way into the quarterly reporting.

Or Facebook could do little things, like thinking more thoughtfully about how the reshare button should be used. The vast majority of reshared content was fine, but the farther you went down a chain of reshares passed from person to person, the more likely you were to find misinformation, hate speech, or violence-inciting content. Facebook's documents contained an intriguing opportunity for someone dedicated to freedom of speech. Imagine a situation in which Alice writes something, her friend Bob reshares it, her friend-of-a-friend Carol reshares it again, and it lands in Dan's news feed. Dan doesn't know Alice; Alice doesn't know Dan. Alice could be an expert, or she could be an agent of an information operation or anything in between. Facebook had run experiments that acknowledged that information gap by graying-out the reshare button for people like Dan: He could still say and share whatever he wanted, but if a piece of content originated more than a friend-of-a-friend away from him, he would have to copy and paste it into the sharing box and start a new sharing chain. For someone who felt that the third-party fact-checking organization was oppressive, a basic change like this could be revolutionary. Graying out the reshare button and asking people to make intentional choices about sharing reduced misinformation the same amount as the entire third-party fact-checking program.

"Which world would you prefer?" I asked the senators. "A world where Dan can say whatever he wants — he can always copy and paste that content and keep spreading it if he wants to — or a world where an unelected committee decides which ideas are dangerous?"

I spoke to the senators about how because Facebook was opaque, the public had never gotten a chance to learn that alternatives existed and that safety did not require censorship. We discussed that what appears on our phones shows that Facebook hasn't changed much since 2008, when the first feed went live. After all, how different can a stream of rectangular posts seem? Yet the reality is that because Facebook had re-architected everything behind the scenes so extensively, we were functionally using a completely different product. We were no longer in our living room having a conversation with our friends and family. Facebook had been trying to find every possible way to tease out our sessions longer in order to pull us away from the real world — from truly meaningful social interactions back to Facebook — and it kept us online longer and longer until we were left in amphitheaters, listening only to a tiny subset of voices chosen not for how much we want to spend time with them, but rather for our willingness to react.

The senators seemed heartened by what they learned. People don't want to be in conflict with each other. They don't want to feel powerless. Committee members wanted to learn more about whether there was a way out, especially when it came to kids, but our time was up. We promised I'd do my best to answer any questions they sent along, and headed on to our next meeting.

I flew home after that second trip exhausted. I had done my tour of duty, and I was very grateful to be back in Puerto Rico. I stayed at home for two very uneventful weeks. Back in August I had been searching for permanent housing after leaving my prior living situation, and I'd interviewed with a couple of San Francisco–style group houses, trying to find a room in a still blistering-hot housing market. I wanted to avoid surprising my potential housemates, so I was honest about what might happen after I came out. There might be paparazzi. There might be angry internet-conspiracists. There might be journalists knocking on the door. I just didn't know. Quite reasonably, none but Alex felt they could provide a room for me.

But none of those consequences played out. Interviewers regularly assume that my life was turned upside down by going public, but I think Puerto Rico's remote location mitigated much if not all of that. I've never been recognized on the streets of San Juan, or if I have, at least people are polite enough not to acknowledge me. I have protective coloration — I'm just another innocuous blonde walking the earth.

Even online, the response was mild. I still have open direct messages on Instagram and Twitter, and an email address posted on my website, yet people don't harass me. No one threatens my family. I really believe this isn't a political issue, and perhaps I did a good enough job presenting it that way that I sidestepped the political battle royal in the United States. The only thing of note during the window immediately after I came out was that one day my new comms team at Bryson Gillette told me we needed to talk about finding ways for more reporters to report on the documents.

Despite all the hiccups in the weeks before the first *Journal* story, we were able to put together a consortium with a small group of European outlets immediately after the paper published Jeff's series. In the days after I came out, press outlets had been peppering Bryson Gillette with more and more questions about me. It was becoming obvious that unless they could report on the Facebook Files themselves, *I'd* become the story. The *Wall Street Journal* didn't want to organize a press consortium. After discussing it with a number of the outlets that were invited, the consensus was that the Associated Press would coordinate a group of English-language journalists. The pool initially expanded to twenty-seven publications in the United States and Europe and eventually grew to a multi-lingual consortium of 156 publications globally. To help them acquire the context they would need to report, every few days we would hold press briefings in which each outlet could ask one or two questions. The information in my disclosures would never have had the impact that it ended up having were it not for the contributions of hundreds of hardworking journalists.

If the United States was now being shaken awake to the reality of Facebook's operations, Europe was years further down that road and had already moved on to discussing what came next. After my two weeks at home I headed for the United Kingdom and Europe for a whistlestop tour of London, Lisbon, Berlin, Brussels, and Paris. Days after I arrived in Europe, Facebook surprised the world by dropping the corporate name they had used for seventeen years and hastily rebranding themselves Meta, loudly proclaiming the future of the company would be in the Metaverse. It felt like the first acknowledgment from the company that they could see they were in trouble.

I feel conflicted about what to call them going forward in this book, because they changed their name to escape the stain of their past wrongdoing, and it feels wrong to leave all their past bad deeds tied to a name that will fade from prominence with time. Even when I was at Facebook, Facebook understood their brand was toxic enough that I never received a single piece of company gear with the company logo on the outside. Yet, for the sake of clarity, I will refer to Meta as Facebook for the remainder of this book.

Facebook had assumed for years that as long as they tried hard to keep the American English-language version of Facebook safe and reasonably presentable, they would escape regulatory scrutiny. That had left the European Union, composed of twenty-seven countries with twenty-four official languages, scrambling to deal with the consequences of a messier and more dangerous version of Facebook in their countries.

When you decide to handle safety on your platform only after the fact, by creating censorship systems to pluck out the bad content, you have to rebuild those systems language by language if you want similar levels of safety elsewhere. You might even have to rewrite them region by region if there are local differences when it comes to what constitutes hate speech or what constitutes "civic" issues. That means finding tens of thousands of examples for each individual tool (one for violence, one for child endangerment, one for bullying, and so forth) for each individual language. Europe suffered the consequences acutely of Facebook judging only languages like English or even German as practical to receive a modicum of safety, while leaving a language with fewer speakers, such as Latvian, out in the cold.

This could have serious consequences. By the time I went public, Facebook for years had been trying conscientiously to find networks glorifying suicide among English-language Instagram users. (And to be clear, the word "suicide" above is not a typo for "suicide prevention".) Yet I've spoken to journalists in Norway who found networks with hundreds of young girls who fetishized self-harm that Facebook did not take down even after the journalists reported them. I can't say for certain why this happened, but I'm guessing that Facebook felt it couldn't justify staffing people to handle suicide on Instagram for "such a small market" like Norway. When the journalists followed up later on, multiple people with accounts within the group of suicide-advocating accounts had died by their own hand. The Europeans had grown tired of Facebook's weak (or nonexistent) attempts to keep their citizens safe and had been talking to academics and researchers for years about what to do. By the time I landed on the other side of the pond, they had a draft piece of legislation called the Digital Services Act.

At the most basic level, there are two ways of writing laws: You can articulate principles or processes you must follow, or you can prohibit certain kinds of behavior. Businesses tend to prefer prohibition-style rulemaking because it's very clear: This is good, that is bad. But the world isn't two-dimensional, especially when it comes to tech. You can't draw simple lines, especially with things that move as fast as technology, to differentiate good from bad behavior.

Sometimes governments will acknowledge how hard it is to describe prohibited features or behavior, and instead focus on prohibited content.

Sometimes those laws ban types of content (like pro-Nazi content in Germany or pro-terrorism content in Australia) or they give individuals the right to demand that platforms remove content about themselves (like child bullying or nonconsensual naked imagery in Australia). One of the dangers of enumerating prohibited content is that lists of what is prohibited tend to grow over time, endangering freedom of speech. But the even larger danger is taking down nonviolating content like counterterrorism materials. When a machine learning classifier is being calibrated to determine "Is this bad?" legal penalties for getting that decision wrong incentivize taking down large amounts of nonviolating content by accident. Even at Facebook, where they strove to bias toward accuracy over comprehensiveness, they found that 75 percent of counterterrorism content in some Arabic dialects was being labeled as terrorism content and removed. This has serious national security implications. The best way to fight terrorism is to have communities deradicalize themselves. Taking down the voices of people arguing against terrorism is how we help terrorism spread.

I've been asked countless times, by everyone from concerned citizens to senators, "Give me the top ten things we should prohibit to fix social media." I always respond, "That's not how you solve a problem like this."

Part of why that won't "solve social media" is that technology is constantly shifting. Many of the ways to "fix" Facebook don't translate to a social media site like TikTok that is even more aggressively personalized and hyper-amplifying. Another problem is that in some ways the "bad" parts of social media aren't the causes of the harm we see out in the world, they are themselves symptoms of the problem. The real problem Facebook represents is that it is fundamentally unaccountable, doesn't disclose its research and data, and no two users encounter the same experience on it. What you see when you open Facebook is different from what I see when I open it. If when we talk we see a common pattern, we have no idea how representative that pattern is. When researchers have tried to help people collaborate to gather data that measures what Facebook is doing, Facebook has threatened to sue them until they stop.

What became clear to me when I testified before the European Parliament was that these legislators were done just asking Facebook to treat them with the same respect they seemed to show authorities in the United States. They saw the horrific effects of the spending imbalances, and when they went to investigate, Facebook wouldn't answer basic questions about what I call "linguistic equity"—the idea that all people deserve safety online no matter what language they speak. It didn't matter whether they were asking how many moderators spoke French or exactly how many fact-checking articles Facebook was paying for in German; EU officials got crickets in response.

So if it's hard to describe prohibited behaviors and prohibiting content is ineffective (or even harmful), what are we going to do? Fortunately, there is another way to hold these companies accountable. A way that does not rely on telling them what can and cannot be said on their platforms, but instead makes the design of their systems safer. The European Union wrote a law called the Digital Services Act (DSA) that focuses not on speech, but on the system that enables and amplifies it. If the core problem with social media platforms is that they are opaque, the DSA attempts to address that by requiring companies like Facebook to draw back the curtains they had pulled tight. Under the DSA, very large platforms—software systems used by more than 10 percent of the European Union in any given month—must complete risk assessments covering a broad range of issues, from freedom of speech and respect for civil liberties to the systematic risks of the algorithms' decisions about what content would be distributed and to whom. The law demanded that those assessments be shared with the EU and the public. If platforms want to sweep an issue under the rug and not include it in their assessment, the EU has the right to demand and compel answers.

The Europeans understood that part of why Facebook got so out of control was that it was never subject to oversight, there was no mandated transparency, and therefore the public didn't know what it didn't know. Facebook controlled all the data about how its platforms performed so that no one could come up with their own questions. No one could grade Facebook's homework but Facebook. They knew that if we had our own measuring stick, we might expect the company to improve in measurable ways—ways that would compel them to invest in safety measures and thus diminish their profits.

Facebook took advantage of our blindness, and they were adept at defusing criticism. In the summer of 2021, the only successful advertiser boycott in the company's history arrived with the Stop Hate for Profit campaign. Advertisers were pulling their advertising dollars because of all the press exposing the hate speech flourishing there. Facebook won them back. The company came out and said, "We solved hate speech" and made a great public show of the pages and accounts it took down. The boycott campaign crumbled without Facebook ever having to produce data about how much hate speech was still flourishing and what fraction of the problem they had managed to address.

One of the most important provisions of the Digital Services Act was its articulation for the first time in history of the right to data for academics and civil society groups. There is no industry in the world with the impact of social media in which only the company gets to know how their products operate and the consequences of the choices made in their corporate offices. The EU was inviting scrutiny, bringing to the table researchers who might help find new paths to better social media.

I like to think of laws like the DSA as nutrition labels. In the United States the government does not tell you what you can put in your mouth at dinnertime, but it does require that food producers provide you with accurate information about what you're eating. You can't have a free market without giving consumers accurate information about the products they are using. If platforms knew they had to be honest and transparent with their consumers about the risks of their products, it would force them to think carefully about those risks before they released products to the public. This model, focusing on design rather than speech, is emerging as the best way of reining in these companies and is being adopted in countries around the world.

My testimony at the European Parliament lasted two and a half hours but was far more grueling than my session in the United States. There were maybe 150 to 200 representatives in the tiered rings of desks that radiated out stadium-style from the witness stand. Questions came at me from four or five people at a time, and I would have ten minutes to respond. I took detailed notes as I listened, often through a simultaneous-translation headset, and then I had to stitch together my responses on the fly. I told someone later this was the largest group I had testified in front of, and they pointed out there were tens if not hundreds of thousands who watched my Senate testimony. "Yes," I replied, "but I couldn't see any of those people." If I thought I was tired when I walked into the Senate, it was nothing compared to my exhaustion when I got back on the plane home at the end of the three weeks. It had been back-to-back involvement, twelve-to-fourteen-hour days, six days a week.

But it was worth it. The following May, exactly a year after I had left Facebook, I was back in Brussels for a speaking event, and I was invited to a special session of the committee I had addressed in the fall. I wasn't entirely sure why I was there. I couldn't imagine why I was needed. The Digital Services Act had passed just a month before. The only details in the briefing for that day said that I was giving fifteen minutes of testimony at the tail end of a series of speakers before me.

One by one the people who had written and marshaled the Digital Services Act through the legislative process over the previous few years got up and talked about what the law would mean for the European Union and the potential role model it could provide for legislation in other countries. That the DSA made explicit that technology lives in democracy's house. But they also explained why I was in the room that day. Again and again they addressed me directly and said they had gotten it over the finish line because of the information in my disclosures. They thanked me.

I don't handle being the center of attention well. I feel very awkward, and praise elicits more self-doubt than self-congratulation. That day was no different, but my heart swelled to see their satisfaction, no, their excitement for a different and better future. This was far more their accomplishment than mine—they had, after all, spent at least four years lining up the dominos to be knocked down. But if they wanted to welcome me into their circle, I would celebrate with them.

By then I had so much more invested in the process. I had listened to the stories of mothers who had lost their children to self-harm; I had talked with adult children who could no longer speak to their parents without fights over conspiracies driving them apart, and to people who felt like there was no possibility of change. Over the course of three tours I had spent six weeks on the road in Europe, meeting with anyone who asked for help to make change real.

A year earlier I had decided that I could not solve the problems I saw at Facebook alone, nor could Facebook solve its own problems alone. The only chance I saw was that we could solve them together. And now I was living in the first moment of that hope.

CHAPTER 15

Onward and Upward

> This is where we are.
> Where do we go from here?
> First, we must massively assert our dignity and worth.
> We must stand up amidst a system that still oppresses us
> and develop an unassailable and majestic sense of values.
>
> —Martin Luther King Jr.

"**W**as it worth it?"

A year after I walked into the Senate hearing chamber and into the public sphere, this was the question reiterated in almost every interview on the anniversary of my emergence as the Facebook whistleblower. Did I have regrets? Would I do it again?

The answer, at least for me, was simple. Nothing feels as good as giving another person hope — you can grind your way up mountains all day every day with that for fuel. I had walked into the Senate with the selfish goal of insulating myself from Facebook's wrath and had walked out into a life I had never imagined before. I now found myself regularly called upon to give people a ladder out of the pit of despair that Facebook had invested so much money in digging.

A year later, my message was the same as the one I had delivered in my opening address to the Senate: "These problems are solvable. A safer, more enjoyable social media experience is possible." I understood what the interviewer really wanted to hear. What are the landmarks that prove the world has changed? What laws were passed? In what ways are the products

Facebook and Instagram different now? What criminal charges were pressed? Who was punished for their lies or for the harm they caused?

We make movies and create epic paintings about the cathartic, distinct moments when the world changed because they are deeply satisfying in their clarity. Unfortunately, that's not how most significant change really works. Change is a process of evolution that begins in the mind of a single person and flows outward until it becomes inevitable. Change feels impossible until it's undeniable, yet no one makes movies about the struggle until it's over.

We like to point to the passage of laws because they're clear symbols of victory, but laws are lagging indicators that culture has changed. First, norms have to be established, then consensus must be reached on what it means to transgress those norms. The journalists interviewing me thought we had long ago reached consensus. Every day they talked to people about the state of technology and its future. Many had actually read through thousands of pages of the Facebook Files. Didn't everyone already know how dangerous social media can be? For them, the lack of a landmark law in the United States was a challenge. If the Facebook Files weren't big enough to get things fixed, what possibly could be?

But as obvious as the revelations in my disclosures may have seemed to those journalists over the previous year, the average person in the United States had still not heard about "the Facebook whistleblower." Or more amusingly, if they *had* heard of a Facebook whistleblower, they thought that person had exposed what Cambridge Analytica had been up to, a scandal four years in the past. I know this because when I'm on planes I'm often asked what I do for work, and at least half the time, whoever I'm talking to has no idea what I'm talking about.

We all try to contextualize new experiences with previous ones. To many tech journalists, because of how omnipresent I was in their reporting in the fall of 2021, I had become Ralph Nader standing up to the car companies. They felt they had watched a watershed moment, and they expected to see Congress do something akin to the 1966 National Traffic and Motor Vehicle Safety Act, which made car manufacturers responsible for the safety of their customers. We take for granted that cars today are as safe as they are, but back then, cars had a fatality rate of 5.7 deaths per 100 million miles driven, or close to *five* times the fatality rate they had in 2019. We have come to assume cars will only get safer, yet in 1965, the fatality rate had been rising for years.

In the 1960s almost no cars had seat belts, which were treated as amenities you had to pay extra to secure. As a result, people were often thrown through their windshields in crashes. People would be decapitated when badly designed glove-compartment doors flew open at just the wrong angle in crashes and passengers were thrown against metal dashboards. Drivers were regularly impaled on their steering wheels. No cars had airbags. We were told that cars were dangerous and we all needed to accept a certain number of deaths if we wanted to be able to drive. Don't worry, the car companies would explain; those who died were responsible for their own deaths because of bad choices. Don't make bad choices like driving fast or driving intoxicated, and you'll be fine.

Then Ralph Nader published *Unsafe at Any Speed: The Designed-In Dangers of the American Automobile* in November 1965. Less than two months later he was testifying in front of Congress about how auto companies knew how to make cars safer but chose not to. He wrote of the decades of research on the power of seat belts to prevent the "second collision" that takes place in an accident when a passenger hits the dashboard (or the ground if thrown from the car). Car companies knew that collapsible steering columns could protect drivers from being impaled, and they knew the new steering columns cost the same as old noncollapsible steering columns. Padded instrument panels are cheaper than chrome or glossy enamel ones and save lives. But do they sell cars?

Nader was clear: The lost lives were not inevitable; they were *intentional choices*. Automotive executives were afraid to be the first to move on safety, reasoning that if they started including safety features or talking about them, people might start asking why *their car* needed safety improvements when no one else was investing in them. The public was outraged to learn that executives worried more about losing sales than about saving lives.

The change the journalists expected to see was embodied in the National Traffic and Motor Vehicle Safety Act, the first comprehensive automobile safety legislation in US history. At the signing, President Lyndon Johnson drew attention to how we can become blind and numb to ongoing tragedy by noting that while 29 American soldiers had died over the recent Labor Day weekend in Vietnam, 614 Americans had perished in automobile accidents. Society was willing to march in the street for 29 fallen soldiers a day, while the automotive dead invisibly piled up. People had been talking about automotive fatalities for decades, but ten months after *Unsafe at Any Speed* was published, suddenly there were new and upgraded vehicle-safety standards. More than that, there was also an agency dedicated to enforcing them and supervising safety recalls when they were transgressed. "Where is our *National Social Media Safety Act?*" was the question journalists really seemed to be asking me. A cynic might look at the lack of a landmark US law and conclude that "This is just proof that nothing can be done."

While it might seem at first that Ralph Nader and I were on a similar journey, that his book and my testimony were equivalent signposts along the same road, it's important to understand how profoundly different cars are from the intangible products that make up our social media algorithms and content delivery systems. The automobile business is radically more transparent than our current social media platforms. Ralph Nader did not do the primary research that enabled him to write *Unsafe at Any Speed*. He was a young lawyer who read through reports from the California Department of Transportation, hundreds of depositions against car manufacturers, patent filings, and even articles published by GM. He had experts to guide him, ranging from disgruntled employees in the industry to auto safety pioneers who had patented the first three-point seat belt. Nader deserves credit for synthesizing this information and communicating it in a way that was able to produce change, but he stood on the shoulders of giants to get off the ground.

Furthermore, Nader was writing for a public that was already aware of the human costs of the automotive industry's underinvestment in safety. People regularly witnessed scenes of bodies being flung through windshields because seat belts were an added, optional expense. They saw reports of people decapitated by poorly designed and weakly latched glove compartment doors. No one doubted that cars killed people; *Unsafe at Any Speed* just pointed out that it didn't have to be this way.

In addition, insurance companies had an economic incentive to fund independent research regarding what "safe enough" ought to mean. Every auto accident that could be blamed on driver negligence was one they didn't have to pay for. Plaintiffs' lawyers had an economic interest in digging through that research and identifying people who had been wrongfully harmed, then forcing car manufacturers to face the consequences in the form of negligence settlements. The automobile industry already had an ecosystem of different actors in which each participant had different strengths and different incentives, but together they all helped pull cars toward the common good.

Sunlight is said to be the best disinfectant because once we can see what problems exist, we're more motivated to fix them. The automobile industry's ecosystem of accountability was able to grow, mature, and diversify only because it could access the sunshine of independent information. Unlike the data centers that host the code that produces our social media experiences, anyone with the money could buy a car, drive it, and if they wanted to, crash it. They could take the car apart. If they didn't know what to look for, they could go to school and earn a master's degree focused on automotive design, in schools that had existed for decades researching the best ways to build cars and trucks.

Today the ecosystem of accountability for social media is still in its infancy. It has been starved of sunshine. The only people who know how social media systems work are the employees at those social media companies, all of whom are vetted carefully before they can access anything sensitive. You can't take a single academic course, anywhere in the world, on the tradeoffs and choices that go into building a social media algorithm or, more importantly, the consequences of those choices.

For more than a decade before I blew the whistle, researchers had begged for access to even basic data, such as "What are the top thousand most viewed posts on Facebook this month?" Facebook kept that information secret for a simple reason — the public would be horrified if it knew the truth. Facebook knew that as long as they kept the curtain closed, as long as we were limited to what we could see on our own screens, they would get to control the narrative regarding how their platforms functioned. If they didn't disclose what choices were being made, we could never ask what the motivations or consequences of those choices were, and

Facebook would be able to define what it meant to "fix" any given harm that might draw attention.

Operating in the dark, Facebook reverted to the 1960s-era automotive industry playbook. The basic message of Nick Clegg's editorial, "It Takes Two to Tango," was one any public relations flack from General Motors would have recognized — "We're not hurting you; your bad choices are hurting you."

"But aren't we now out of the dark?" you might ask me. "Didn't you spend years working on these systems? Didn't you bring thousands of pages of documents forward? Don't we now know what we need to know?"

Over the past year, many people have asked me what are my top five things to fix on Facebook, or what ten things we should ban on social platforms. People I have an immense amount of respect for have insisted that I need to pick one or two things and just push to fix them. They think I needed to accept that that was all I could realistically expect to accomplish. We already know lots of ways to make social media safe. If you and I sat down tomorrow, we could talk for hours about strategies for fixing things, but I feel strongly that if we don't have a community of experts, or even a community of concerned and informed citizens, who can effectively weigh the trade-offs implicit in any intervention, all we will have accomplished is swapping one unaccountable leader for another.

So how do we go about activating that community of concerned citizens?

Any plan to move forward that's premised on me personally proposing the solution is a plan that's doomed to fall short. The "problem" with social media is not a specific feature or a set of design choices. The larger problem is that Facebook is allowed to operate in the dark. When it comes to social media, we aren't just missing the equivalent of an education in automotive design, we're missing an understanding by the public or even academics of why we need the equivalent of seat belts. To be clear — this isn't just Facebook's fault. Before Facebook there were Myspace and Friendster. For twenty years social platforms have hid basic information about their systems so we'd just have to trust them.

It doesn't have to be this way. The European Parliament's Digital

Services Act is one major step forward. It acknowledges that we don't yet know what the rules of the road should be, beyond greater transparency. That we need a different, more respectful relationship between these mega-platforms and the public, grounded in the truth. At its heart, the DSA is attempting to raise awareness so we can begin to discuss what to do next. With the DSA, at least European academics, civil-society groups, and public officials are entitled to ask questions, via independent researchers. No one can be expected to catch up with what the platforms already know about their systems if those platforms don't disclose the harms they're already aware of. With the DSA, Europe now is entitled to learn about those harms through mandatory risk reporting.

The DSA is a great first step, but it doesn't yet ensure the level of transparency needed to allow the ecosystem of accountability to flourish. We need public streams of social media data that allow academics, civil-society groups, and even individual concerned parents to keep tabs on whether what's coming out of the public relations departments of social media companies is actually aligned with the way those companies are operating. That data can be privacy-protected—no individual identities need be exposed—but the underlying data must be made widely available. Believe it or not, YouTube influencers discussing social media data are part of how we will democratically govern Facebook.

Now that you have read this book, it may not seem like it, but you also are a meaningful part of the solution. One of the most critical factors that drove down our national automotive fatality rate was concerned citizens like Mothers Against Drunk Driving who kept safety at the forefront of the conversation. I hope this book has given you enough context to feel comfortable demanding more, because your voice is important for how we democratically govern social platforms in the future.

One of the projects I want to work on over the next couple of years is teaching a million people the "physics" of social media. We should never give raw, large-scale, real social media data to college or high school students, but we can build simulated social networks that mirror the repeating patterns of real-life social networks. With that, we could teach everything from messy, sprawling big-data data-science classes to introductory ones for high school students. When Ralph Nader published *Unsafe at*

Any Speed, there were tens if not hundreds of thousands of professional automotive engineers with mature professional organizations dating back sixty years who could assess, confirm, and vouch for his allegations. When the first wave of Facebook reporting began, there were maybe hundreds of people who deeply understood the information in the disclosures, many of whom were still employed by social media companies. What would social media be like if we expected hundreds of thousands of people to hold accountable the information environment of the twenty-first century?

Opening up data from the social networks allows us to stop relying on Big Tech to grade their own homework. Critical tools available to the ecosystem of accountability include boycotts and stock divestments. If the platforms won't negotiate with the ecosystem, the ecosystem always has the power to withdraw its economic support. People often say social media boycotts don't work, but part of why they don't work is that we can't measure whether social platforms are making progress on the issues we care about. In the summer of 2020, when the Black Lives Matter protests spread to major cities across the United States, a coalition of advertisers set up the #StopHateForProfit advertiser boycott, the most extensive such action to date. Facebook was able to snuff out the boycott by claiming they had made huge leaps forward in managing hate speech on their platform. Yet documents in my disclosures from a year later said Facebook was only taking down 3 to 5 percent of hate speech on Facebook.

Similarly, no major coordinated divestment campaign has been launched against the social media giants. I've spoken with managers at endowments, sovereign wealth funds, and major pension funds. They insist that they want to act but claim they can't take a stand until norms are established. People managing money can't divest without a clear portrait of what "good enough" looks like.

Over the past year I've met with traditional investors like large mutual funds, and they consistently repeat how surprised they were by Facebook's plummeting stock price and how this impacted their portfolios. Facebook was supposed to be a safe bet that always went up. Because they didn't know what data to demand so they could assess where the company stood with regard to risk, or even what questions to ask in order to assess what the underlying risks to that business were, they couldn't put pressure on

Facebook to build a safer product. I guarantee you that the largest share-holders of the major car companies have closed-door meetings discussing how car safety and litigation risk are managed. What decisions might Facebook have made differently if they'd had to engage with external actors in a similar way?

Let's imagine that tomorrow the top story we saw on our phones was that all the social platforms were opening up their data to independent researchers. Would that be enough to push us toward the social media we deserve? I believe the root of all change is the belief that change itself is possible and desirable. Part of what keeps many from demanding action from social media companies like Facebook or TikTok is that until the very recent passage of the Digital Services Act, the only governments we saw stepping up to hold Big Tech accountable were authoritarian states like China. It's hard to demand change when the role models you can point to seem to validate Facebook's propaganda that safety can come only at the cost of free speech.

Unlike the United States or even Europe, China acted immediately when concerns began to surface about social media's impact on children's mental health. Beyond deploying large numbers of moderators to delete information about eating disorders or self-harm, Douyin, the Chinese ana-log to TikTok, has a "youth mode" that allows people under eighteen to use the app only for at most *forty minutes* a day and locks them out from 10 p.m. to 6 a.m. every night. People who pay attention to technology policy look at approaches like this and rightly see the paternalism and control at their core. These policies presume that kids can't be trusted to decide when to stop using these apps and that there are topics kids should not be allowed to access. While time limits and subject blocklists might be effective for addressing compulsive use and exposure to genuinely harmful topics, these strategies represent a slippery slope that could lead to platforms (or the government) deciding what topics people should be allowed to learn about or even how people can access social media in the first place.

China has done this by binding its national identification cards to all social media accounts so that they can reliably identify every person under eighteen. Right now in the United States, if you lose your social media account, you can simply register another one. Not so in China: Once your identification card/passport is banned, you're banned for good. The same technology and moderators that keep children from content promoting self-harm are also used to block any ideas or people viewed as challenging those in power. It's an utterly authoritarian and paternalistic approach. But other approaches are possible. In the Ralph Nader example, he was able to point to epidemiological studies of the comparative safety of wearing a seat belt versus not wearing one. This gave the public a place to focus their demands for improvements to cars: How can we reduce the impact of the "second collision"? Why not look at similar data from social media? What are the equivalent fulcrums?

The two biggest factors I would put forth as being equally important to the second collision in cars are *unintentional usage* and *algorithmic escalation*. Unintentional or undesired usage is any usage of a digital product beyond what you genuinely want. You might call it *regretted usage*. Algorithmic escalation is the process by which a recommender system progressively pushes you toward more extreme information. It's almost never intentionally part of the design of the system but naturally occurs because the computer systems that direct our attention only optimize for getting more clicks out of us, and we're attracted to more extreme content, even if we don't consciously like it. People think they're opting into one kind of experience, and like the boiling frog, transition into another. Both of these are by-products of the business models of social media companies. Right now advertising-funded platforms are incentivized to make their products as "sticky" as possible—every minute more they keep you online is another minute when you might view an ad or click on it and make them money. They don't capture how much regret you feel for the time you spend on them or the costs you might pay for that time. If part of what drives that stickiness is that you're pushed to more extreme content, that algorithmic bias is an invisible cost to the companies.

Right now we put up with this because we tell ourselves we're lucky to get "free" products like Facebook or Instagram. Many have bought into the industry narrative that it's our own fault if we become addicted to them. Meanwhile in many communities we've gutted our in-person opportunities to connect. People go to church less often, they join fraternal orders and other civic groups at lower rates, and we collectively invest less in shared infrastructure like community centers and green spaces where we might connect. The very people most often harmed by social media are the same people who lack many (or any) options for in-person community.

Let's take a moment and consider an alternative form of social media that's based on an alternative set of *values*. If digital social spaces may be the only social spaces for some people, we must value human beings online the same way we value them in person. What if our new digital code of conduct for companies were grounded in them demonstrating dignity and respect? When a social media platform prioritizes how long they can keep us online over the net value of that time to each of us, they treat us as a means to an end. We're reduced from being humans bearing real costs to mere conduits for profit. What if we demanded that tech companies demonstrate that they value our dignity as human beings — and promote our autonomy when using their services as proof? How would these products look different?

We've known for twenty years that if you make a product function a little slower, people will use it way less. Imagine if social platforms asked us at noon, when we still had willpower, at what time we wanted to go to bed in the evening (or at least, when we wanted to stop using their products). At Google, we would obsess over tiny time delays (latency) because we could see in the experimental dashboards that usage dropped off a cliff when the site got slower. Facebook could easily let children pick their bedtimes on Instagram, and then for two hours before that deadline make the app progressively a little slower and a little slower — so gradually it wasn't obvious — until you got tired and turned in.

Think of how many kids go to school every day hungover on Instagram and what the lifelong costs will be from that sleep deprivation. Those costs borne by our children are optional. I know this because that slowdown code is already live on Instagram. If you're a hacker and you're downloading content off Instagram, the platform won't delete your account. After all, you can easily just make a new one. Instead, they will slow your account down more and more over time. If Instagram cared about insomnia in our children and the multitude of risk factors it exposes children to, they could

326

stop it tomorrow by giving children more control over their usage. But they choose not to. Every doomscrolling fourteen-year-old brings in revenue. Launching a feature like that draws attention to the compulsion these products can provoke. Might some people question their own usage when presented with a choice like that? If Instagram moves first and TikTok doesn't follow, will they give up market share? It's almost as if you can hear the echoes of the world we had before the 1966 National Traffic and Motor Vehicle Safety Act, when the lack of an established floor for automotive safety left each car company afraid to make the first move to save lives.

When it comes to algorithmic escalation, we have a similar fork in the road to navigate. We could take the Chinese strategy (or the strategy used by Facebook today) and delete the "bad" content, but denying people choices doesn't respect the autonomy of individuals. How might we approach algorithmic amplification if we were trying to value dignity and autonomy? Imagine a child or an adult being led from "healthy eating" to "eating disorder" content. We know that without intervening, Instagram will continue to show more extreme content over time. How could we respect the dignity and autonomy of that child or adult by putting them back in the driver's seat? Instagram today has something called a topic model that assigns at least one subject/category to every piece of content on Instagram. Instagram has run experiments in the past in which they take those clusters of content and they poll users by asking, "When you see posts like this, how does it make you feel?"

Imagine if Instagram began to notice when someone's feed began to be populated by more and more content that other users said made them feel bad about their bodies or otherwise depressed. Respecting their autonomy would mean recognizing the role of Instagram's algorithms in boosting that content and placing questions under those posts that read, "Do you want to keep seeing content like this?" or "Would you like to see less?" The algorithms only escalate because we let them.

On an even simpler level, imagine if we gave people the right to have algorithmic products *forget* what they've learned about us so we can start from scratch. When I was recovering from my hospitalization, I watched a great deal of dark content on Netflix. At some point I got better enough that I started to be turned off by the TV shows and movies Netflix always seemed to recommend when I opened the app. Netflix hadn't figured out that I had moved on.

In my case, being shown stale recommendations tied to my past wasn't a big deal — I moved in with some friends who already had Netflix, and I was shocked at how much more positive and fun their version of Netflix was. When it comes to children and Instagram, I've been told by child psychologists that people recovering from eating disorders talk about disordered eating content following them around on Instagram. It doesn't matter if they're making all the right choices in their daily life; unless they're willing to sacrifice their memories and friends, and risk being socially isolated, they must place themselves in harm's way with exposure to content that could trigger their eating disorders. We're forcing children to choose between their past and their future. We don't have to — but Instagram chooses to, because they're afraid to admit that their products are dangerous. They're afraid they'll lose advertising dollars when people stay on the site just slightly less after their hyper-personalized algorithms reset. Remember, this is an industry where your stock price can plummet if a core metric like revenue misses expectations by 1 percent.

These are only a couple of strategies to illustrate how we can optimize for freedom and safety at the same time. We have a large tool chest, and we'll invent more tools the more people get a chance to learn about how these systems operate. But platforms won't use them unless they feel pressure to do so. Facebook knew its platforms were causing harm. Its stock price continued to soar because nobody else knew it. The company avoided spending resources on the safety systems they needed, artificially plumping up their profits. They had many users who wouldn't have opened their app every day if they'd known the truth of what they were using. Facebook happily reported revenue from advertisers who wouldn't have accepted the brand risk if they knew about the harms of the platform. Facebook's profits were contingent on no one knowing how large the gap between Facebook's and Instagram's public narratives and the truth had grown. When that gap was small, the risk of exposure was small. As the gap grew, the need to protect against the truth grew larger and larger.

Corporations are facing a fundamental decision over the next few decades. We are entering an era in which employees understand they're disposable, while also understanding that the systems they work on, the systems that fuel our economy, are more opaque to those outside the corporate walls than they were in the past. The next generation of employees understands in a very different way than preceding generations that if they don't act to share information with the public about black-box systems that run on chips and data centers, the public won't get the information it needs to provide oversight to tech companies, with potentially deadly consequences. The future is likely to bring many more Frances Haugens.

Companies will have to decide whether they want to be transparent with the public, and therefore not have to fear their employees, or whether they want to police their employees and accept that they'll lose the best talent. People don't like to have to keep secrets. People don't like to lie. Neither activity is free. And the benefits of lying are in fact long-term liabilities.

Facebook didn't want to see their lies as liabilities. When the truth broke, they lost users and advertisers, and had to dramatically increase their safety spending. Twice in the last five years, first in 2018 and then again in 2021, Facebook/Meta set the record for the largest one-day value drop in stock market history, both times when their narratives met reality head-on. We should expect to see this over and over again with opaque companies in the future. *Lies are liabilities.* Truth is the foundation of long-term success.

Acknowledgments

If I had fully grasped what writing a book entails when I began a little over a year ago, I would have likely chosen other projects to embark upon. Thankfully I have been supported unflaggingly by my husband, Alex, through all the unimaginable curveballs that came with this project. Without him, this book would not have been possible.

This book was also made possible on a much longer time horizon by the string of "moms" I have accumulated throughout my life. Thank you to Jan Bohnsack who shielded me and pushed me forward throughout my elementary years. Thank you to Linda Muhly for reminding me I could want more out of my future, and to Kate Hamm for guiding me to being a young adult. Thank you to Maureen Taylor who taught me the liberating power of forgiving others and the importance of claiming agency in my own life. Thank you to Libby Liu (or, my "lawyer mom"), who looked after me as the chaos swirled when I came out. One day I will remember to bring my dress shoes with me when I testify.

And most importantly, thank you to my biological mom, Alice, another person without whom this book would not have been possible. She has provided endless assistance in shaping and refining this book, and I could not have made it through the last two months without an incredibly heavy lift on her part. One day I hope I get the opportunity to tell the story of how you taught me the importance of taking responsibility and never giving up until things are made right.

Thank you to Taylor Owen and Helen Hayes for their critique of this manuscript and their suggestions for how to enrich it. It is unquestionably more polished and persuasive because of your efforts. Thank you to Marshal and Alex for being my technical readers.

ACKNOWLEDGMENTS

Thank you to Emily Schwartz and Bill Burton who have helped me refine and find my voice all the way from the intense last-minute sessions right before I filmed *60 Minutes* till today. The press consortium and the overall rollout would not have been possible without them. Thank you to Reset for arranging the opportunities to work with leaders across Europe, which provided my crash course education in public policy along with many once-in-a-lifetime moments. Thank you particularly to Dylan for the countless twelve-to-fourteen-hour days on the road with me.

Thank you to Matthew Gray for providing the space and coaching for me to fall in love with programming. Thank you to Dan Clancy and Jack Menzel for teaching me about information retrieval and the wonders of search quality. I think it's unlikely I would have gotten to go on the many adventures I have over the past fifteen years if you two hadn't coached me so extensively.

Thank you to Angie and Annie, Tobias, Steve, Lee, Antoun, Leslie, Ed, Brooke, and Toby who never gave up on me while I was literally getting back on my feet.

Thank you to James Steyer and Common Sense Media for your guidance and support as I have found my voice to push for accountability and transparency with big tech. Thank you to Frank McCourt and the McCourt Institute for believing in my team and me and making our advocacy possible.

Lastly, thank you to the legal team that has supported me along the way: Mary Inman, Sarah Alexander, Lawrence Lessig, Andrew Bakaj, Marcia Hofmann, Brian Klein, Liana Chen, Kronenberger Rosenfeld.

Index

INDEX

INDEX

INDEX

About the Author

FRANCES HAUGEN is an American data scientist, algorithmic product manager, and whistleblower. A graduate of the Franklin W. Olin College of Engineering and Harvard Business School, she worked at Google, Yelp, and Pinterest before joining Facebook/Meta in 2019 to work in its Civic Integrity department. In 2021 she disclosed tens of thousands of pages of internal documents to the Securities and Exchange Commission and the *Wall Street Journal,* revealing Facebook's awareness of and complicity in radicalization and political violence around the world.